Introduction to the Theory of Atomic and Molecular Collisions

Introduction to the Theory of Atomic and Molecular Collisions

J. N. Murrell and S. D. Bosanac

School of Chemistry and Molecular Sciences
The University of Sussex, UK

JOHN WILEY & SONS
Chichester • New York • Brisbane • Toronto • Singapore

Other Wiley Editorial Offices

John Wiley & Sons, Inc., 605 Third Avenue,
New York, NY 10158-0012, USA

Jacaranda Wiley Ltd, G.P.O. Box 859, Brisbane,
Queensland 4001, Australia

John Wiley & Sons (Canada) Ltd, 22 Worcester Road,
Rexdale, Ontario M9W 1L1, Canada

John Wiley & Sons (SEA) Pte Ltd, 37 Jalan Pemimpin 05-04,
Block B, Union Industrial Building, Singapore 2057

Library of Congress Cataloging-in-Publication Data:

Murrell, J. N. (John Norman)
Introduction to the theory of atomic and molecular collisions

Bibliography: p.
Includes index.
1. Collisions (Nuclear physics) 2. Scattering (Physics) I. Bosanac, Slobodan D. II. Title.
QC794.6.C6M87 1989 539.7′57 89-9017
ISBN 0 471 92365 6

British Library Cataloguing in Publication Data:

Introduction to the theory of atomic and molecular collisions.
1. Atoms. Collisions.
I. Murrell, J. N. (John Norman)
II. Bosanac, S. D.
539.7′54

ISBN 0 471 92365 6

Phototypesetting by Thomson Press (India) Limited, New Delhi.
Printed in Great Britain by Biddles Ltd., Guildford, Surrey.

Contents

Preface

An understanding of processes that can occur in collisions between atoms and molecules is central to many branches of physics and chemistry. In such collisions there can be transfer of momentum, of energy, of atoms or of electrons. The most direct method of studying molecular collisions is by a crossed beam experiment. Scattering theory provides the connection between the results of such experiments and the potentials of the interacting molecules. The theory involves quite difficult mathematical techniques but it also encompasses many important concepts which can be illuminated by simple models. The collision of hard shapes, for example, can explain some of the features of energy transfer. However, if we want a quantitative comparison between the results of theory and experiment this will usually require extensive computational analysis which, even today, can test the power of the largest computer.

Scattering theory is still a very active subject. Only for the elastic collisions of atoms is there a good understanding of the underlying theory and efficient numerical methods for computation. Collisions which exchange energy but not atoms present problems, but perhaps not too many that are fundamental. However, it is still only for rather simple systems, such as atom–diatomic molecule collisions, that extensive calculations can be found in the literature. For collisions that lead to an exchange of atoms, there is currently a great deal of discussion about the best theoretical approach, and very few accurate numerical results are yet available.

The basic tools at our disposal are classical and quantum mechanics, and the classical approach can be considerably enhanced, or the quantum simplified, by semi-classical theory. Classical mechanics cannot give the exact answer for a well-resolved experiment; atoms and molecules are, after all, quantum systems. However, it gives a very good qualitative picture, because the de Broglie wavelengths of the particles are frequently similar to, or shorter than, the size of the molecule. Moreover, the concepts that enter classical mechanics are generally familiar and not irrelevant to an understanding of the quantum mechanical problem.

This book provides an introduction to the theory of low energy collisions. These are typically those occurring with velocities similar to those in room temperature gases—say collisions at 0.1 to 1 eV rather than 1 keV or 1 MeV. An understanding of this theory should be expected now of anyone intending to do research in gas phase spectroscopy, or reactivity, or in many other branches of physics and chemistry. We have aimed for a balance between the classical and quantum approaches and a balance between the four main branches of the subject: elastic, inelastic and reactive scattering and electron excitation. We have, however, not covered collisions of electrons or photons with molecules because these topics take us into new theoretical techniques which are rather outside the rest of the field. The main emphasis of the book is on understanding but the computational techniques have not been ignored.

We are extremely grateful to the British Council for providing travel funds which enabled us to meet on several occasions to produce this book.

We are very grateful for the assistance we received from Charlotte Cundall and Liana Reić in the preparation of the manuscript.

JOHN MURRELL, School of Molecular Sciences, University of Sussex
DANKO BOSANAC, Institute Rugjer Boskovic, Zagreb
1988

Chapter 1

Introduction to scattering theory

1.1 The crossed beam experiment

Many physical and chemical properties of gases, liquids and solids can be understood from the potentials that act between the molecules. If we know the form of these potentials, i.e. how they depend on the distance between the molecules and on the relative orientations of the molecules, then we can use statistical mechanics to calculate the properties of these macroscopic phases.

In some exceptionally simple cases data from an experiment on a macroscopic sample can be analysed to deduce the form of the potential; the (P, V, T) data for an inert gas is such a case. In general, however, statistical mechanics acts as a valve on the route between a macroscopic property and the potential; it is usually possible to calculate the property from the potential but impossible to calculate the potential from the property. For this reason potentials usually have to be deduced from other sources. They can, in principle, be calculated directly from the equations of quantum mechanics, but in practice this can only be realized for systems with a few light atoms.

The most direct experimental method of determining intermolecular potentials is by examining the results of individual atomic or molecular collisions, and this is usually achieved by a crossed beam scattering experiment. Two beams of atoms or molecules with a narrow spread of velocities are crossed in a collision chamber and the scattered atoms or molecules are detected at various angles to the incident beams. The density in the collision chamber is too low for multiple collision events to be significant so that the scattering pattern can be analysed according to the theory of single collisions; this will be discussed in Chapter 2.

Atom–atom collisions normally involve only an exchange of momenta between the atoms, and this is called an elastic collision. In some circumstances, which usually require high beam energies, the electronic state of one or both atoms is changed in the collision, or an electron is transferred between the atoms. These are called inelastic processes because they involve a transfer of energy between the translational motion and internal states (electronic states) of the atoms. This transfer is also referred to as a non-adiabatic process because the electronic state of the atom–atom pair is changed in the collision, which implies that the system has made a transition between two potential energy curves. We will examine the details of this process in Chapter 6.

When atoms are in collision with molecules, or molecules with other molecules, we can observe inelastic processes in which energy is transferred between translational motion and the internal rotational and vibrational states of the molecules. This energy transfer is responsible for many important processes that occur in gases and liquids such as the broadening of rotation–vibration spectral lines and, most important, the spontaneous movement of a macroscopic phase towards a condition of equilibrium. Some information about the intermolecular potential responsible for this energy transfer can also be deduced from spectroscopic studies of the gas, but more detailed information can be obtained from the analysis of molecular beam scattering data.

To obtain maximum information from a crossed beam experiment one needs to prepare the beam molecules in specific internal states (rotational and vibrational) as well as to have discrete translational energies. Any lack of specificity of the beam conditions will supress the wave-character of the scattering process. These scattered molecules are also produced in a variety of internal states and the angular distribution of these should, ideally, be determined separately.

Atoms and molecules in collision can also produce chemical reactions. From macroscopic measurements of reaction rates one extracts energy-averaged reaction rate parameters such as activation energies and pre-exponential factors. By studying reactions under beam conditions with state selection of the constituents one obtains collision cross-sections that are functions of the impact energy and the internal states of the reactants. Also, one can be certain that the reactions occur by single collision processes which is not always clear from macroscopic experiments.

We have emphasized at the beginning of this section that it is difficult to obtain detailed information about an intermolecular potential by analysing experimental data on a macroscopic sample and that the main objective of the beam experiment is to deduce such potentials. We must not, however, give the impression that it is easy to deduce potentials from beam data. The link between an experimental observation and a potential is also made by a theory, which is called scattering theory, and, as for macroscopic properties, it is much easier to deduce the property from the potential than the potential from the property. This means that for most intermolecular potentials we are forced to deduce its accuracy by seeing whether the scattering data we deduce from it agree with the experimental results or not.

The ideal molecular beam experiment produces a large amount of data which is very sensitive to the intermolecular potential. It is still very early days in respect to the knowledge about potentials that has so far been acquired, but there is no doubt that there is a growing body of knowledge to be acquired from experiments of the future.

1.2 Waves and particles

Few of us have the skill at snooker or pool to make the balls travel precisely where we want them to, but most of us have an understanding of the dynamics of hard sphere collisions which is sufficient to make the game enjoyable. It would be a much more difficult game with soft balls or elliptical balls but the trajectories would still follow the laws of classical mechanics and we could at least play the game on a computer. The type of calculations we would have to do are quite similar to those that are needed when we wish to predict by classical mechanics what happens when atoms and molecules collide with one another.

Figure 1.2 illustrates diffraction and interference of waves. The pattern in this case has no obvious relationship to the movement of the snooker balls in Figure 1.1 and yet this picture also has relevance to atom and molecular collisions. If these two widely different pictures represent the same collision process, what is the true representation?

There is no easy answer to this question because by such pictures we are trying to comprehend something which we cannot directly experience, using our knowledge about the macroscopic world which we think we understand. It so happens that atoms and molecules sometimes behave as if they were snooker balls and sometimes as if they were waves; this is the wave–particle duality of atomic systems. What we observe from experiments involving atomic and molecular collisions are such things as the angular distribution of the particles after collision, the fraction of chemical reactions that have occurred, the number

Figure 1.1 Hard sphere collisions

Figure 1.2 Model of waves breaking around Napier (N.Z.) harbour. (This picture was kindly provided by Hydraulics Research Ltd, Wallingford, UK)

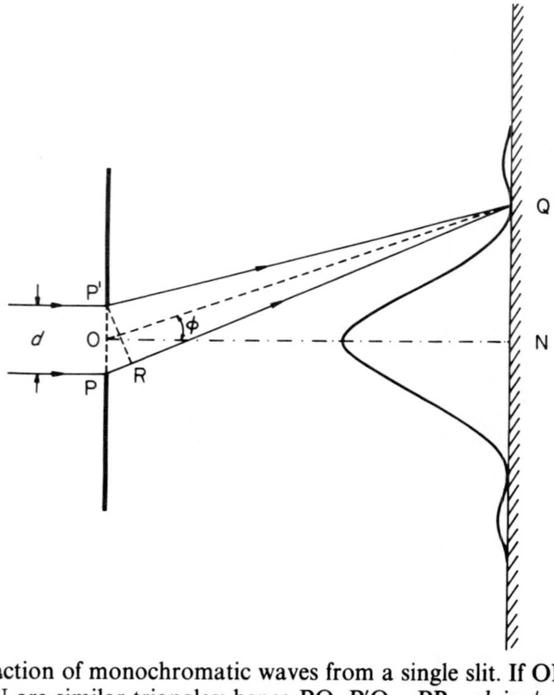

Figure 1.3 Diffraction of monochromatic waves from a single slit. If ON $\gg d$, PP′R and OQN are similar triangles; hence PQ–P′Q = PR = $d \sin \phi = \lambda/2$

of systems in excited energy states, etc. It is these features that we attempt to explain and predict using the mathematical techniques of classical and wave mechanics and that can be interpreted through the pictures we have shown. Not surprisingly some of the terminology which we will use is taken directly from those analogues.

The most important distinction between wave and classical particle behaviour is that waves exhibit interference. This is manifested in various ways but broadly speaking by spatial oscillations whose separation depends on the wavelength of the wave. Figure 1.3, for example, shows the characteristic interference pattern arising from the passage of monochromatic light of wavelength λ through a single slit of width d.

This pattern can be attributed to the interference of two spherical waves originating at the opposite sides of the slit, and therefore the angle subtended at the centre of the slit by the first minimum is

$$\phi = \sin^{-1}\left(\frac{\lambda}{2d}\right) \simeq \frac{\lambda}{2d} \tag{1.1}$$

Of course, detection of this wavelike behaviour is only possible if the angular resolution of the detector is better than ϕ.[†]

In quantum mechanics the wavelength of a particle with momentum p is given by the de Broglie relationship

$$\lambda = \frac{h}{p} = \frac{h}{(2mE)^{1/2}} \tag{1.2}$$

where h is Planck's constant, m is the mass and E the kinetic energy. The detection of wave behaviour in the scattering of atoms is therefore most easy with light particles at low energies. A hydrogen atom with 1 eV (1.6×10^{-19} J) of kinetic energy has, for example, a wavelength of approximately 0.3 Å. Chlorine atoms at the same energy have a wavelength six times shorter.

As a model for the interference effects arising from the collision of two atoms we can replace the slit in Figure 1.3 by a potential whose width is approximately equal to the *minimum* distance between the centres of the two atoms which pass each other without measurable deflection. For neutral atoms this is typically in the range 1–10 Å. The scattering of particles from different parts of this potential will then manifest itself as an interference pattern, and we can deduce from 1.1 that to see a wavelike interference pattern from the collision of 1 eV hydrogen atoms with some other atom (say a potential of range 3 Å) requires a detector with angular resolution better than 0.05 radians (3°). For chlorine atoms at the same energy the resolution needs to be better than 0.5°.

Of course, there will be other factors in an experiment that limit resolution other than the properties of the detector. For example, if the atoms undergoing collisions do not all have the same energy then there will be a spread of

[†] A full treatment of the single slit diffraction experiment, which explains the shapes and intensities of the interference peaks, can be found in M. Born and E. Wolf, *Principles of Optics*, 4th ed., Pergamon Press, 1970.

wavelengths in the beam and this will smear out the interference pattern—for the same reason that a spread of wavelengths in the slit experiment will smear out the interference pattern.

We hope it is clear from the above discussion that we are not questioning the existence of wave character in the scattering of atomic particles; that is always present. What is at issue is whether the interference features can be resolved. If they cannot be resolved then we may anticipte that the classical picture of atomic scattering is usually adequate to explain the observations.

1.3 Trajectories, wave packets and stationary states

In classical mechanics the state of a system is described by the set of coordinates, q_i, and conjugate momenta, p_i, of the particles. If the state changes with time then this can be followed from the time variation of these variables, $q_i(t)$, $p_i(t)$, which can be plotted as a curve in (q, p) space which we call a trajectory. If the system is isolated then its energy and momenta are conserved and the variables evolve according to the equations of classical mechanics of conserved systems. These are most conveniently expressed in Hamilton's form

$$\frac{dq_i}{dt} = \frac{\partial H}{\partial p_i}, \qquad \frac{dp_i}{dt} = -\frac{\partial H}{\partial q_i} \tag{1.3}$$

where H is the classical hamiltonian function which is the functional expression for the total energy expressed in the variables q and p. For example, a particle of mass m moving in a one-dimensional potential $V(q)$ and having linear momentum p has a hamiltonian function

$$H = \frac{p^2}{2m} + V(q) \tag{1.4}$$

so that Hamilton's equations in this case are

$$\frac{dq}{dt} = \frac{p}{m}, \qquad \frac{dp}{dt} = -\frac{dV(q)}{dq} \tag{1.5}$$

The methods employed to integrate Hamilton's equations and to deduce the trajectories will be described in Chapter 5.

However, the coordinates and momenta of atomic particles cannot be simultaneously determined with unlimited precision. The product of the uncertainties in our knowledge of the instantaneous values of these quantities is limited by Heisenberg's uncertainty principle

$$\Delta q \cdot \Delta p > h \tag{1.6}$$

The distribution of coordinates of an atomic system is represented by a probability function $P(q)$ which in quantum theory is given by the square modulus of a wave function $\Psi(q)$:

$$P(q) = |\Psi(q)|^2 \tag{1.7}$$

If $P(q)$ does not change with time then the system is said to be in a stationary state and $\Psi(q)$ is a solution of the Schrodinger equation

$$\mathbf{H}\Psi(q) = E\Psi(q) \tag{1.8}$$

where $\mathbf{H}$ is the quantum mechanical hamiltonian and E is the total energy. The quantum mechanical operator for the system described by the classical hamiltonian (1.4) is

$$\mathbf{H} = \frac{-\hbar^2}{2m}\frac{\mathrm{d}^2}{\mathrm{d}q^2} + V(q) \tag{1.9}$$

If $P(q)$ changes with time the system is not in a stationary state and its wave function evolves according to the *time-dependent* Schrodinger equation

$$-\frac{\hbar}{i}\frac{\partial\Psi(q,t)}{\partial t} = \mathbf{H}\Psi(q,t) \tag{1.10}$$

A localized probability function which changes with time is called a wave packet. The simplest example is a gaussian distribution, which in one dimension is

$$P(q) = \frac{1}{a\pi^{1/2}}\exp\left(\frac{-q^2}{a^2}\right) \tag{1.11}$$

This has a half-width of 1.665 a (for simplicity one often calls a the width of the wave packet).

To find the time evolution of a wave packet one must calculate, from (1.10), the time evolution of the wave function. For the initial gaussian distribution (1.11) the most general wave function is

$$\Psi(t=0) = \frac{1}{\pi^{1/4}a^{1/2}}\exp\left(\frac{-q^2}{2a^2}\right)\exp(\mathrm{i}k_0 q) \tag{1.12}$$

which we see contains the phase factor $\exp(\mathrm{i}k_0 q)$ which is not determined from (1.7): k_0 is called the wave vector.

The average value of the momentum in the wave packet [which is given by the expectation value of the momentum operator $(-\mathrm{i}\hbar\partial/\partial q)$] is $\hbar k_0$ and hence the classical interpretation of this wave function is that it represents the probability pulse of a particle, centred initially at $q = 0$, moving in the direction of increasing q with momentum $p = \hbar k_0$. Moreover, it can be shown that the width of the momentum distribution is $\hbar/a$ so that the product of the widths of the position and momentum distributions satisfies the uncertainty principle (1.6).

Unfortunately (1.10) can be solved in closed form only for a few special cases, and even for the simplest case, with the particles moving in a constant potential, the exact solution is quite complicated. A wave packet evolves with time by moving its centre at a velocity $k_0\hbar/m$, the classical average velocity of the particles. In fact, it can be shown that the movement of the centre of the wave packet follows the classical equations of motion, which is known as Ehrenfest's theorem. The wave packet will also spread and, if the potential is not constant, will change its shape and even split into several wave packets. However, the width and shape are fairly stable (i.e. change more slowly than the movement of the centre of the wave packet) if the initial width, a, is much larger than the de Broglie wavelength of the average momentum, $\lambda_0 = 2\pi/k_0$. This condition is generally satisfied in molecular beam experiments because of the high momentum resolution (narrow spread of translational energy) of the beams and the comparatively long time of the pulses of particles in the beams.

If the characteristic time for spreading of a wave packet is much longer than the time which the wave packet spends in the interaction region of colliding molecules then for all practical purposes we can replace the gaussian wave packet (1.12) by a so-called plane wave

$$\Psi = \exp(ik_0 q) \quad 1.13$$

which represents a continuous stream of particle with definite momentum $\hbar k_0$ moving in the direction q. In this way we have replaced a rather complicated time-dependent problem by a much simpler time-independent problem, and analytical solutions for the scattering process can be obtained in some cases.

To illustrate this we examine the problem of a beam of particles meeting a potential step and being partly reflected and partly transmitted. The model is one-dimensional and the parameters are defined in Figure 1.4. The Schrodinger equation for this system is, from (1.8) and (1.9),

$$\left[\frac{-\hbar^2}{2m}\frac{d^2}{dq^2} + (V - E)\right]\Psi = 0 \quad 1.14$$

and in any region in which V is independent of q this can be integrated to give particular solutions:

$$\begin{aligned}\Psi_{\pm} &= \exp\left\{\pm\frac{i}{\hbar}[2m(E - V)]^{1/2}q\right\} \\ &= \exp\left(\pm\frac{ipq}{\hbar}\right) = \exp(\pm ikq) \quad 1.15\end{aligned}$$

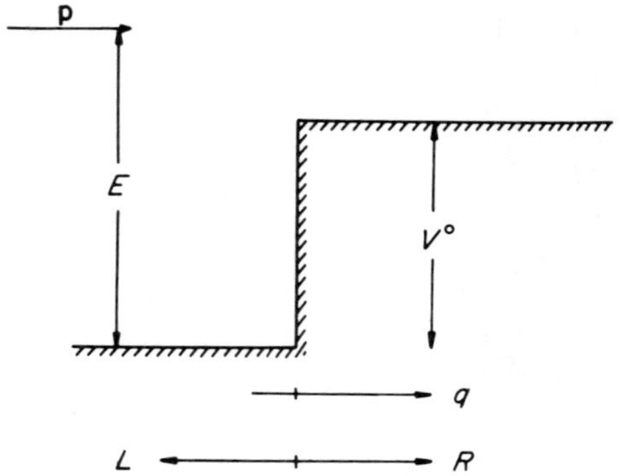

Figure 1.4 Particle scattering on a one-dimensional stepped potential

The wave function Ψ_+ represents a particle moving with momentum p in the direction of increasing q and Ψ_- represents a particle moving in the opposite direction.

The general solution of (1.14) in a region of constant potential is a linear combination of the particular solutions Ψ_+ and Ψ_-. To the left of the step ($q < 0$) we have

$$\Psi_L = A_L e^{ik_1 q} + B_L e^{-ik_1 q} \tag{1.16}$$

where

$$k_1 = \frac{(2mE)^{1/2}}{\hbar} \tag{1.17}$$

To the right of the step ($q > 0$) we have

$$\Psi_R = A_R e^{ik_2 q} + B_R e^{-ik_2 q} \tag{1.18}$$

where

$$k_2 = \frac{[2m(E - V^0)]^{1/2}}{\hbar} \tag{1.19}$$

We can now apply the necessary conditions of continuity on Ψ and on $d\Psi/dq$ at the boundary $q = 0$. This leads to the two equations

$$A_L + B_L = A_R + B_R \tag{1.20}$$

and

$$k_1(A_L - B_L) = k_2(A_R - B_R) \tag{1.21}$$

We now impose a further condition, from the nature of the experiment, which is that particles are sent towards the step from the left and none are sent in from the right (which would be a different experiment). We therefore take B_R to be zero and can now solve (1.20) and (1.21) to obtain

$$\frac{B_L}{A_L} = \frac{k_1 - k_2}{k_1 + k_2} \tag{1.22}$$

and

$$\frac{A_R}{A_L} = \frac{2k_1}{k_1 + k_2} \tag{1.23}$$

The probability densities of the incident, reflected and transmitted beams are

$$|A_L e^{ik_2 q}|^2 = |A_L|^2 \tag{1.24}$$

$$|B_L e^{-ik_2 q}|^2 = |B_L|^2 \tag{1.25}$$

and

$$|A_R e^{ik_2 q}|^2 = |A_R|^2 \tag{1.26}$$

respectively. The number of particles passing a point in unit time, which is called the flux of particles, is proportional to the product of the probability density and the momentum. The ratio of the flux in the reflected beam to the flux in the incident beam, which we call the reflectance, R, is therefore

$$R = \frac{k_1 |B_L|^2}{k_1 |A_L|^2} = \left(\frac{k_1 - k_2}{k_1 + k_2}\right)^2 \qquad 1.27$$

and the ratio of the flux in the transmitted beam to the flux in the incident beam, which we call the transmittance, T, is

$$T = \frac{k_2 |A_R|^2}{k_1 |A_L|^2} = \frac{4k_1 k_2}{(k_1 + k_2)^2} \qquad 1.28$$

As the transmitted beam has a different translational energy to the incident beam it represents an inelastic scattering event†. It can be confirmed from (1.27) and (1.28) that

$$T + R = 1 \qquad 1.29$$

Hence the solutions we have obtained conserve flux.

If the energy of the incident beam, E, is just equal to the height of the step, V^0, then $k_2 = 0$ and the transmittance becomes zero. For energies below this ($E < V^0$), k_2 is imaginary and although there is some tunnelling into the step there is no transmittance. Figure 1.5 shows how the transmittance changes smoothly as the energy in the incident beam increases above the step height. This can be contrasted with the classical result in which T changes discontinuously from zero to one at $E = V^0$. We shall see in later chapters that this is a characteristic

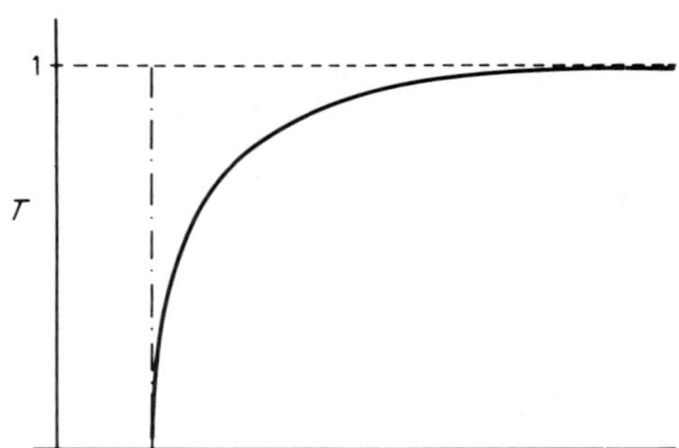

Figure 1.5 Transmission coefficient for scattering from the potential of Figure 1.4. The quantum result (1.28) is the full line and the classical result is the stepped function

†An interesting feature of this result provides a warning about using model potentials having a steplike behaviour. Both the reflectance and transmittance are independent of the mass of the particle, so that they are finite even in the case when m is infinite. However, in this limit the quantum system should approach the classical limit $R = 0$ and $T = 1$ but the quantum results clearly fail to do this. The reason for this failure is that the step in the potential appears as a rapidly varying function of q, no matter how large k_0 is, i.e. no matter how large the mass of the particle. If the step in the potential is represented by a smoothly varying function of q, then a value of k_0 can be found for which the incident wave varies more rapidly than the potential, and in such a case the reflectance correctly goes to zero when the mass of the particle becomes infinite.

difference between classical and quantum behaviour; sharp changes which occur in classical mechanics are smoothed out in quantum mechanics.

The example we have just given shows how time-independent quantum mechanics can be used to describe a scattering process. We stress the simplicity of the mathematics and contrast this with the complexity that arises in the equivalent time-dependent description of a wave packet travelling from the left and being partly reflected and partly transmitted; we must refer interested readers to an original source for this problem.†

1.4 Semi-classical theory

We remind the reader that the first successful treatment of quantized energy levels in atomic systems was the Bohr theory of the hydrogen atom spectrum. It is common today to think of this theory as thoroughly outdated and conceptually much less satisfactory than modern quantum mechanics which replaced it. However, Bohr theory held its place in physics for nearly twenty years and that shows that it had some notable successes. The proper view of Bohr theory is that it fills a valuable intermediate position between classical and quantum mechanics which is referred to as semi-classical mechanics. We shall see that it plays an important role in scattering theory by introducing quantal features, notably interference, into classical mechanics. One cannot say that classical mechanics is always mathematically simpler than quantum mechanics, although, as we shall see later, it is currently easier to apply to many-dimensional systems. The big advantage it has is that its results are usually simpler to understand from our knowledge of the behaviour of macroscopic bodies. For this reason semi-classical mechanics allows us to extend what is already a good conceptual base.

From classical mechanics we derive directly the probabilities for scattering events. For example, if we wish to calculate the probability that the collision of an atom and a molecule leads to a chemical reaction then we can examine a large number of collisions between the two by calculating classical trajectories for different approach angles. We then find the fraction of the total that leads to a reaction.

In quantum mechanics we calculate wave functions and determine probabilities from the square moduli of these (1.7). If there are several component functions that contribute to the total wave function then it is the square of the sum of these that gives the probability, and it is the cross terms in this that gives rise to interference effects: *quantum mechanical probabilities are given by the squares of sums NOT by the sum of squares (probabilities).*

In semi-classical mechanics one combines the concept of a classical trajectory with the interference effect from quantum theory. This is done by associating a phase ϕ with a trajectory, which has the value

$$\phi = \hbar^{-1} \int_{q_i}^{q_f} p \, dq \tag{1.30}$$

†M. H. Bramhall and B. M. Casper, *Am. J. Phys.*, **38**, 1136 (1970).

where the integral is taken from the initial to the final point on the trajectory, and is called the classical action.†

If there are several trajectories which connect certain initial and final states the phases will generally be different and the probability of making a transition between these states is obtained from the squared sums of the amplitudes

$$P = \left| \sum_{\mu} P_{\mu}^{1/2} e^{i\phi_{\mu}} \right|^2 \qquad 1.31$$

where P_{μ} is the probability of the system having the trajectory μ and ϕ_{μ} is the phase for that trajectory.

The proof of (1.30), which we give in Appendix 1, is obtained by examining the classical limit of quantum mechanics ($h \to 0$) in the so-called JWKB approximation. It is clear from (1.31) that if there is only one classical trajectory that connects specified initial and final states then the classical and semi-classical probabilities are equal. If there is more than one then interference arises from the terms in (1.31), such as

$$(P_{\alpha}P_{\beta})^{1/2}(e^{i(\phi_{\alpha}-\phi_{\beta})} + e^{-i(\phi_{\alpha}-\phi_{\beta})}) = 2(P_{\alpha}P_{\beta})^{1/2}\cos(\phi_{\alpha}-\phi_{\beta}) \qquad 1.32$$

There is a second use of semi-classical techniques in scattering theory, which has its origin in the Bohr formula for finding the special classical trajectories (orbits) amongst all possible classical trajectories that are associated with the quantum bound states of the system. For orbital motion in one dimension this recipe is

$$\oint p\, dq = nh \qquad 1.33$$

where the integral is taken over a complete orbit and n is an integer. As well as giving the exact energy levels of the hydrogen atom this recipe also gives the energy levels of the harmonic oscillator if n is replaced by $(n + \frac{1}{2})$, to allow for zero point energy, and of the diatomic rigid rotor if n is replaced by $[n(n + 1)]^{1/2}$. Thus the Bohr quantization condition with these amendments can be used in classical trajectory calculations to identify the rotation–vibration quantum states of diatomic molecules which are found as the initial and final states of the scattering process.

The failure of Bohr theory was that there was no generally satisfactory and easily used replacement for (1.33) for multi-dimensional problems in which the variables are not separable. For this reason there was no satisfactory Bohr theory of any molecular electronic state, not even for H_2^+, and likewise the rotation–vibration states of polyatomic molecules cannot be identified with specific trajectories. Although there have been later developments in this field the problem of the semi-classical quantization of multi-dimensional systems has not been fully solved.

†There is a theorem in classical mechanics called the theorem of least action which states that for a conservative system the trajectory $[q(t), p(t)]$ that leads from the initial to the final state is the path for which the action is stationary with respect to any small variation of the path.

1.5 Laboratory and centre-of-mass coordinates

The coordinates of the centre of mass are constant in a conserved system; this is a consequence of the condition that the linear momentum of the system is a constant of the motion. In scattering theory one therefore usually works with equations in which the centre-of-mass variables have been separated from the internal coordinates. The results obtained from a scattering calculation therefore need some further analysis before they can be compared with data from a crossed beam experiment which are obviously most simply described in coordinates that refer to the structure of the apparatus. A typical apparatus is shown in Figure 1.6. In other words, we need to transform results obtained in a scattering coordinate or centre-of-mass system (C) to the laboratory coordinate system (L). The transformation from one coordinate system to another is quite straightforward but what is less straightforward is to find the relationship between the scattering angle in the two systems.

Suppose that in the L system we have particles with masses m_1 and m_2 colliding with velocity vectors $\mathbf{v}_1$ and $\mathbf{v}_2$. In most experiments $\mathbf{v}_1$ and $\mathbf{v}_2$ are perpendicular to one another. Their relative velocity is clearly

$$\mathbf{v} = \mathbf{v}_1 - \mathbf{v}_2 \tag{1.34}$$

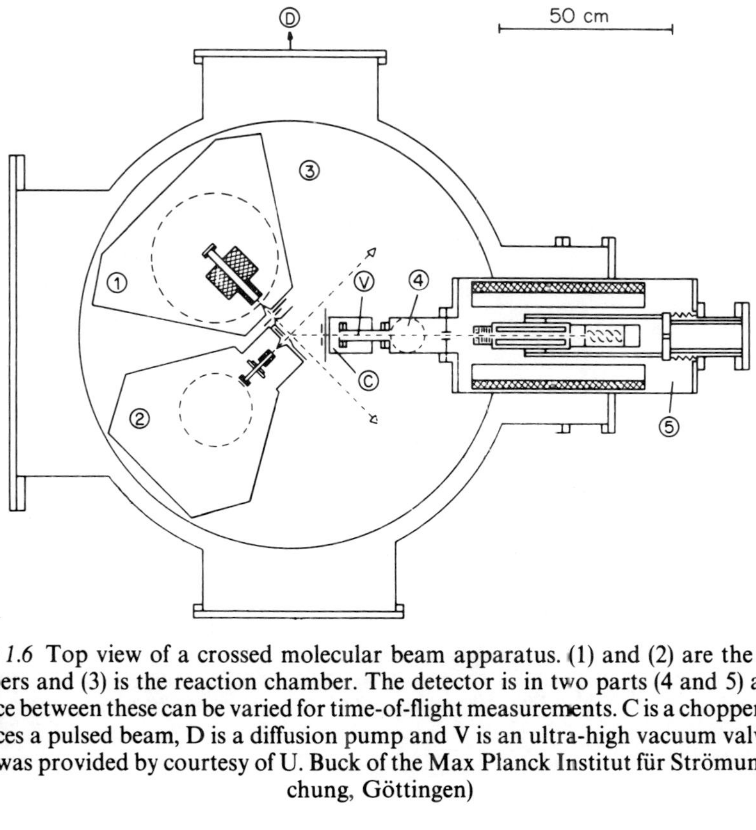

Figure 1.6 Top view of a crossed molecular beam apparatus. (1) and (2) are the source chambers and (3) is the reaction chamber. The detector is in two parts (4 and 5) and the distance between these can be varied for time-of-flight measurements. C is a chopper which produces a pulsed beam, D is a diffusion pump and V is an ultra-high vacuum valve (The figure was provided by courtesy of U. Buck of the Max Planck Institut für Strömungsforschung, Göttingen)

and the velocity of their centre of mass is

$$\mathbf{V} = \frac{1}{M}(m_1\mathbf{v}_1 + m_2\mathbf{v}_2) \qquad 1.35$$

where $M = m_1 + m_2$.† $\mathbf{V}$ and $\mathbf{v}$ are the velocities in the C coordinate system and it is easy to confirm that the kinetic energy in the two systems is the same:

$$2T = m_1 v_1^2 + m_2 v_2^2 = MV^2 + \mu v^2 \qquad 1.36$$

where μ is the reduced mass

$$\mu = \frac{m_1 m_2}{M} \qquad 1.37$$

The relationship between the L and C velocities is illustrated in Figure 1.7 which is called a Newton diagram. The centre-of-mass point P divides $\mathbf{v}$ in inverse ratios of the particle masses.

After the collision we have particles with masses m_1' and m_2' moving with velocity vectors in the L system, $\mathbf{v}_1'$ and $\mathbf{v}_2'$ respectively, and their relative velocity vector is

$$\mathbf{v}' = \mathbf{v}_1' - \mathbf{v}_2' \qquad 1.38$$

In the C system the velocity of the centre of mass is unchanged; hence, by comparison with (1.35),

$$\mathbf{V} = \frac{1}{M}(m_1'\mathbf{v}_1' + m_2'\mathbf{v}_2') \qquad 1.39$$

where $m_1 + m_2 = m_1' + m_2'$.

The treatment is sufficiently general so far to include elastic, inelastic and reactive collisions. However, we will first examine the simplest case, which is elastic scattering, because from the conservation of energy we know that the length of $\mathbf{v}$ is unchanged by the collision. Also P is a fixed point (because V is

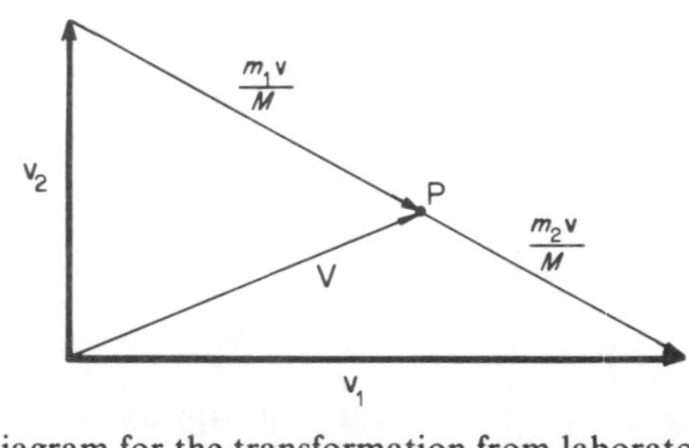

Figure 1.7 Newton diagram for the transformation from laboratory (v_1, v_2) to centre-of-mass velocities (V, v). The laboratory beams have been taken as perpendicular

†The position of the centre of mass is clearly $\mathbf{R} = (m_1\mathbf{r}_1 + m_2\mathbf{r}_2)/M$, and (1.35) is the rate of change of this expression.

unchanged) and the ratio in which P divides $\mathbf{v}$ is also unchanged (because the masses are constant or, equivalently, to conserve momentum). Thus the ends of the vector $\mathbf{v}'$ lie on circles with a common centre P as shown in Figure 1.8. Usually the laboratory scattering angle is measured for just one of the particles (the primary beam), which we take to be particle 1. If $m_1 < m_2$ a laboratory measurement of θ_L uniquely determines the centre-of-mass scattering angle θ_C (Figure 1.8a). If $m_1 > m_2$ (Figure 1.8b) then there are two possible values of θ_C for each value of θ_L.

For inelastic scattering energy is transferred between internal states of the colliding particles and the translational motion, so that the magnitude of $\mathbf{v}'$ may be greater or less than that of $\mathbf{v}$. However, the ratio in which P divides the $\mathbf{v}'$ vector is unchanged because the masses of the particles are unchanged. In reactive scattering the masses may also change and then both the magnitude of $\mathbf{v}'$ and the ratio of division by P is different from that of $\mathbf{v}$. In this most general case the radii of the circles on which $\mathbf{v}_1'$ and $\mathbf{v}_2'$ lie must be determined by the condition that energy and momentum are conserved in the collision.

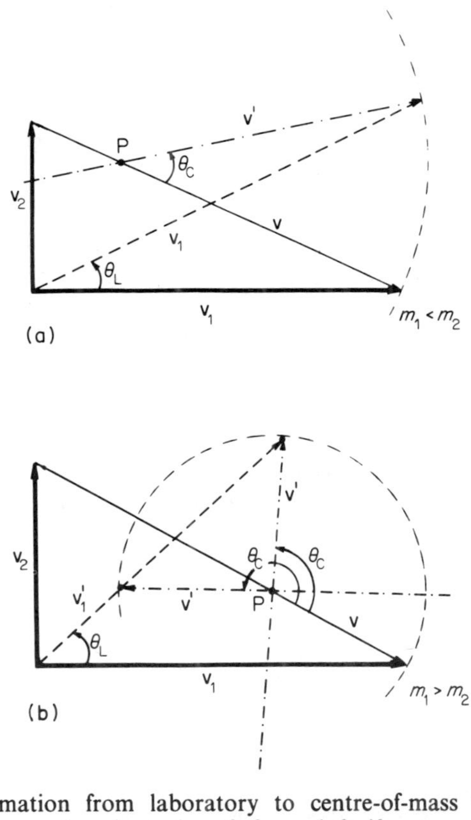

Figure 1.8 Transformation from laboratory to centre-of-mass velocities for elastic scattering. When $m_1 < m_2$ there is a unique θ_C for each θ_L; if $m_1 > m_2$ there are two possible values of θ_C for each θ_L

For energy conservation we have

$$\tfrac{1}{2}\mu \mathbf{v}^2 + E_i = \tfrac{1}{2}\mu' \mathbf{v}'^2 + E_f \qquad 1.40$$

where E_i and E_f are the initial and final internal energies respectively. The magnitude of $\mathbf{v}'$ can be determined from this equation. The point P then divides $\mathbf{v}'$ in the ratio of m_1' and m_2', to conserve momentum. Figure 1.9 illustrates a case in which the internal energy is increased by the collision ($\mathbf{v}' < \mathbf{v}$), the reactants have equal masses and the products have masses in the ratio 1:3.

For elastic scattering a measurement of the intensity of scattered particles as a function of the scattering angle and the translational energy of the particles provides complete information about the system. For inelastic and reactive collisions we need in addition to measure the change in translational energy after the collision. This is usually achieved by a time-of-flight experiment on a chopped crossed beam. The time of arrival of particles at the detector is measured for each scattering angle and from this the velocities of particles after collision can be determined. Knowing $\mathbf{v}_1'$ (say), $\mathbf{v}_2'$ is determined by the conservation of momentum and hence $E_i - E_f$ is determined from equation (1.40).

1.6 Summary of systems to be examined

We have three theories of the scattering process at our disposal, classical, semi-classical and quantum, which provide, in that order, an increasingly satisfactory description but which are increasingly difficult to apply. We have three principal types of scattering problem, elástic, inelastic and reactive, which, in that order, are increasingly difficult to solve.

Collisions between atoms and molecules are multi-dimensional problems and apart from a few models of elastic scattering the equations must be solved numerically. Our success in using both classical and quantum theory has increased enormously in recent years with the advent of large-memory and fast computers and with advances in the techniques of numerical analysis.

For elastic scattering we have no difficulty in applying all three forms of theory and a comparison between them allows us to see the circumstances in which the classical or semi-classical methods are satisfactory. The simplest examples of inelastic scattering are the collision of an atom with a diatomic molecule treated

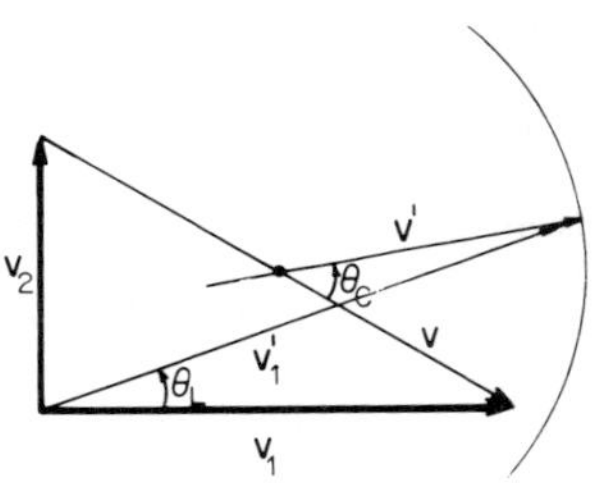

Figure 1.9 Transformation from laboratory to centre-of-mass velocities for reactive scattering: $v' < v$, $m_1 = m_2$, $m_2' = 3m_1'$

as a rigid rotor and the collinear collision of an atom and a diatomic molecule. These can also be treated by all three mechanics, and we can again make useful comparisons which should be generally valid for the important topic of the transfer of energy between the translational, rotational and vibrational states of molecules in collisions.

In classical mechanics the time taken to integrate the classical equations, and hence to calculate a trajectory, increases rapidly with the number of variables; typically as the square or cube depending on the algorithm used. One saves computer time by finding variables which are constants of the motion, such as the position of the centre of mass and the angular momentum variables, but this may be at the cost of introducing complications in the remaining equations, and the gain is sometimes marginal. We will examine this point in a later chapter when we examine the classical trajectory method in more detail.

The time-independent Schrodinger equation is a second-order partial differential equation and it is not currently feasible to integrate such an equation numerically except in one dimension.† Thus, multi-dimensional quantum mechanical problems, whether they are for bound states or the continuum states needed in scattering theory, are always solved by writing the wave function as an expansion in known functions which is referred to as the basis. For bound states these expansion functions usually cover all the dimensions of the wave function (e.g. the use of atomic orbitals in the molecular orbital or valence bond approach to electronic bound states). In scattering theory the bases usually provide functions of one dimension less than the total wave function, the last dimension being the distance between the centres of mass of the colliding systems; in this dimension the wave function is integrated numerically. The choice of variables, and the subsequent choice of basis functions in these variables, is crucial to success, and it is, for example, the difficulty of finding good bases for reactive systems that has limited the achievements of quantum mechanics in this field. The quantum mechanical approach to scattering is currently limited to relatively simple systems, as will be illustrated in later chapters.

One of the important techniques for solving bound state problems in quantum mechanics is perturbation theory. This relies for its success on the fact that the wave functions of the perturbed system are quite close to those of the unperturbed system. The equivalent in scattering theory might seem to be that the probability for inelastic events is small. However, this criterion fails in many circumstances because the meaning of smallness is different in bound-state and scattering calculations. For bound-state calculations it is sufficient if the corrections are small relative to the separations of unperturbed energy levels. In scattering calculations the transition probabilities must be small in absolute magnitude, which is much more difficult to achieve. To illustrate this difference we consider an expansion series

$$S = 100 + 10 + 1 + 0.1 + \cdots\cdots \qquad 1.41$$

†Feasible is perhaps too strong a word, but even in two dimensions numerical integration is very demanding of computer time.

from which it would appear that after the third term a high accuracy is achieved. This is true if we require only the relative accuracy, but if S represents a phase then even the last term in S is not sufficient to give an accurate representation of, say, $\sin(S)$.

It can be anticipated that any situation in which classical trajectories are strongly curved will give transition probabilites that are not accurately given by quantum mechanical perturbation theory, and generally there are always trajectories of this type for collisions of atoms and molecules. We will examine the perturbation equations of scattering theory, which are known as the Born equations, in Chapter 3.

Chapter 2

Elastic scattering

2.1 Classical trajectories for the central force problem

An elastic collision is one in which the internal energies of the colliding species do not change. The most important, and theoretically simplest, example of this is the collision of two atoms without changes in their electronic states. Elastic scattering also occurs in molecular collisions but is usually of secondary interest to the inelastic events in which there is a change in molecular rotational and vibrational energies. However, in collisions between a light atom and a heavy molecule, for example, there is very little energy transfer and the scattering will be dominated by elastic collisions.

In the centre-of-mass coordinate system described in Section 1.5, the elastic collision of two atoms produces a change in the direction of the relative velocity vector $\mathbf{v}$ but not a change in its magnitude. The scattering is wholly determined by the change in the orientation of $\mathbf{v}$ and therefore the two-body collision may be formally reduced to the motion of a single particle with reduced mass μ (1.37) in a potential $V(\mathbf{r})$, where $\mathbf{r}$ in the interparticle vector. For spherical particles, such as atoms, the potential only depends on their separation $V(r)$, and this is called a central force problem.

Because the angular momentum of the system

$$\mathbf{L} = \mu\mathbf{r} \times \mathbf{v} \qquad 2.1$$

is a constant of the motion, and because the forces perpendicular to $\mathbf{r}$ are zero in a central force potential, the motion of the particle is always confined to the plane defined by $\mathbf{v}$ and $\mathbf{r}$. In such a case the classical equations of motion are most conveniently expressed in polar coordinates (r, ϑ) which are defined in Figure 2.1. The kinetic energy in these coordinates is given by†

$$T = \tfrac{1}{2}\mu\dot{\mathbf{r}}^2 = \tfrac{1}{2}\mu(\dot{z}^2 + \dot{x}^2) = \tfrac{1}{2}\mu(\dot{r}^2 + r^2\dot{\vartheta}^2) \qquad 2.2$$

and the magnitude of the angular momentum is

$$L = \mu r^2\dot{\vartheta} \qquad 2.3$$

†Write $z = r\cos\vartheta$, $x = r\sin\vartheta$ and take time derivatives.

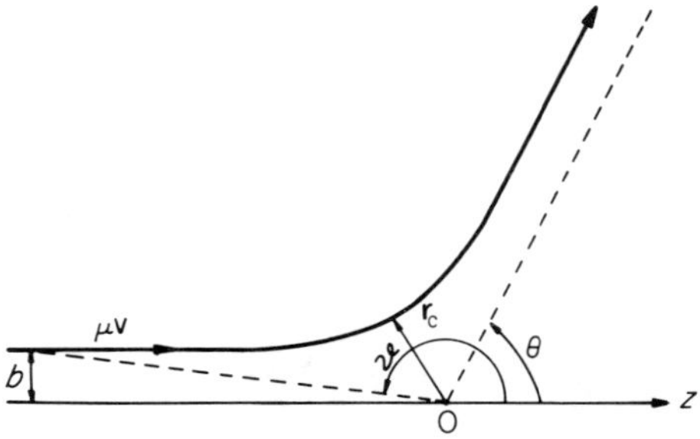

Figure 2.1 Parameters for central field scattering

The total energy is the sum of the kinetic and potential energies

$$E = \tfrac{1}{2}\mu(\dot{r}^2 + r^2\dot{\vartheta}^2) + V(r) \qquad 2.4$$

and replacing $\dot{\vartheta}$ by L and using (2.3) we have

$$E = \tfrac{1}{2}\mu\dot{r}^2 + \frac{L^2}{2\mu r^2} + V(r) \qquad 2.5$$

where $L^2/(2\mu r^2)$ is called the centrifugal potential. It acts with respect to motion in the variable r as an L-dependent repulsive potential, and therefore for each value of L there is an effective potential in the variable r which is

$$V_L(r) = V(r) + \frac{L^2}{2\mu r^2} \qquad 2.6$$

Figure 2.2 shows the potential for the collision of two neon atoms and the effective potentials for different values of L. For $L < 55$ the effective potential possesses a minimum, but this region can only be reached classically if the collision energy is greater than the height of the outer turning point, which is called the centrifugal barrier. The influence of this feature of the effective potential on the scattering angle will be examined later.

Equations (2.3) and (2.5) can be solved to obtain the time variation of r and ϑ, with E and L being constants of the motion. However, if we are interested only in the equation of the trajectory (i.e. the relationship between r and ϑ) rather than the dependence of both on the time, and this is usually the case, then we can eliminate the time from these equations by writing (2.3) as

$$\dot{\vartheta} = \frac{L}{\mu r^2} \qquad 2.7$$

and (2.5) as

$$\dot{r} = \pm\left\{\frac{2[E - V_L(r)]}{\mu}\right\}^{1/2} \qquad 2.8$$

whence

$$\frac{d\vartheta}{dr} = \frac{\dot{\vartheta}}{\dot{r}} = \pm\frac{L}{\mu r^2\{2[E - V_L(r)]/\mu\}^{1/2}} \qquad 2.9$$

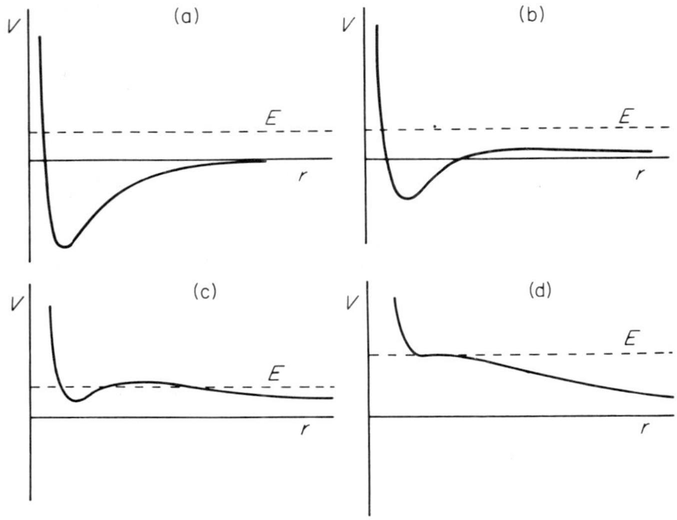

Figure 2.2 Effective potentials (2.6) for atom–atom scattering. (a) $L=0$; (b), (c), (d) $L \neq 0$. (a) and (b) have one classical turning point, (c) has three turning points and (d) shows an inflection point ($L=55$)

The positive sign in these expressions applies to the *inward* part of the trajectory (r decreasing as ϑ decreases) and the negative sign for the outward part of the trajectory (r increasing as ϑ decreases). At the distance of closest approach of the atoms (minimum value of r), E is equal to $V_L(r)$ and $\dot{r}$ is zero. This is called the classical turning point ($r=r_C$).

If we integrate (2.9) from $r=\infty$ to the classical turning point and then back to infinity, which represents the motion of the particle, then we obtain finally a value of ϑ which is the angle into which the particle is scattered. This is called the deflection angle; note that it is a function of L and E. Its value is given by

$$\theta(E, L) = \pi - 2\int_{r_C}^{\infty} \frac{L\,dr}{\mu r^2\{2[E - V_L(r)]/\mu\}^{1/2}} \qquad 2.10$$

where π comes from the initial value of ϑ, i.e. the particle was initially far away in the negative z direction (see Figure 2.1). Both the incoming and outgoing parts of the trajectory give the same contribution to the total integral, and this leads to the factor of two in front of the integral sign.

It is convenient to introduce in (2.10) the parameter b, called the impact parameter, which is defined in Figure 2.1 and which is related to L by the expression

$$L = b\mu v = b(2\mu E)^{1/2} \qquad 2.11$$

Replacing L by b in (2.6) gives the effective potential in the form

$$V_L(r) = V(r) + \frac{b^2 E}{r^2} \qquad 2.12$$

After substituting (2.11) and (2.12) into (2.10) we obtain the following expression for the deflection angle:

$$\theta(E, b) = \pi - 2b \int_{r_C}^{\infty} \frac{dr}{r^2\{1 - (b^2/r^2) - [V(r)/E]\}^{1/2}} \qquad 2.13$$

where we note that the classical turning point, r_C, is a function of E and b.

Analytic solutions of this integral can be found for some simple potentials, as we show shortly, but of more general importance is the general relationship between θ and b which can be deduced from qualitative reasoning.

Firstly, it is obvious that $\theta = \pi$ when $b = 0$. This corresponds to a head-on collision of the two atoms for which there is a rebound along the line of approach arising from the repulsive part of the potential energy curve. It is also obvious that θ will approach zero as b approaches infinity as the atoms will then pass each other without deflection. Finally, if $V(r)$ is a wholly repulsive potential, i.e. one which increases monotonically as r decreases, then θ will decrease monotonically from π to zero as b increases from zero to infinity.

For the collision of hard spheres the potential is zero when r is greater than the sum of the hard sphere radii (d) and infinite for r less than this. The classical turning point is equal to d for an impact parameter b less than d (i.e. r_C is independent of the energy) and equal to b if $b > d$. Since $V(r)$ is zero over the accessible region of the trajectory we have, from (2.13),

$$\theta(b \leqslant d) = \pi - 2b \int_{d}^{\infty} \frac{dr}{r^2(1 - b^2/r^2)^{1/2}} \qquad 2.14$$

Change of variable $u = 1/r$ converts the integral into a standard form from which we obtain

$$\theta(b \leqslant d) = \pi - 2\sin^{-1}\left(\frac{b}{d}\right) \qquad 2.15$$

which we see is independent of the energy.

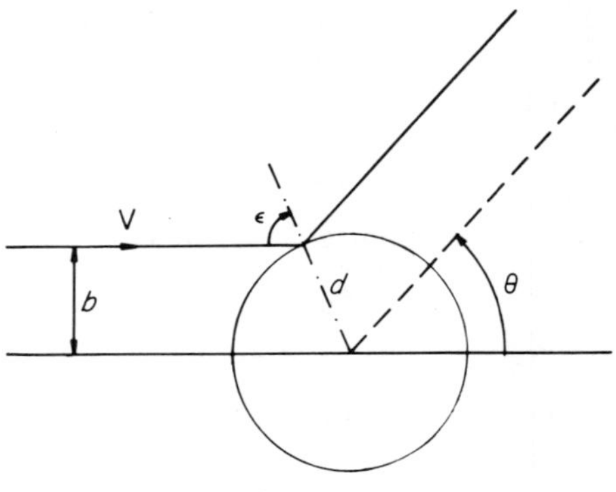

Figure 2.3 Hard sphere scattering

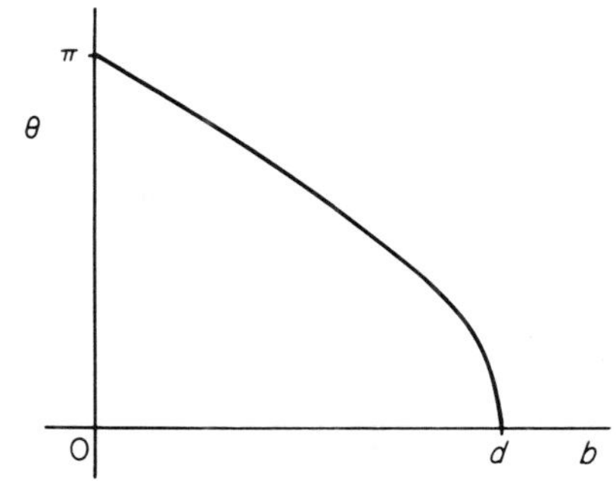

Figure 2.4 Deflection function for hard sphere scattering

For $b > d$ the hard spheres do not collide; hence $\theta = 0$. This result can also be easily deduced from the geometrical construction shown in Figure 2.3, noting that $\theta = \pi - 2\varepsilon$.

Expression (2.15) can be transformed to

$$b = d \sin\left(\frac{\pi - \theta}{2}\right) = d \cos\left(\frac{\theta}{2}\right) \qquad 2.16$$

and this function is sketched in Figure 2.4. Note that θ has a discontinuous slope at $b = d$, which is a particular feature of a hard sphere potential.

The integral in expression (2.13) can also be solved analytically for the repulsive coulomb potential, a result which was first obtained by Rutherford for its application to the scattering of α particles by heavy nuclei. Substituting the function

$$V(r) = \frac{B}{r}, \qquad B > 0 \qquad 2.17$$

into (2.13) we obtain an integral which can be put into standard form ($u = 1/r$)

$$\theta(E, b) = \pi - 2b \int_0^{u_C} \frac{du}{(1 - BE^{-1}u - b^2u^2)^{1/2}} \qquad 2.18$$

where $u_C = r_C^{-1}$ is the positive root of the quadratic expression in the denominator. The required result is

$$\theta(E, b) = 2 \sin^{-1}\left[\frac{BE^{-1}}{(B^2E^{-2} + 4b^2)^{1/2}}\right] \qquad 2.19$$

and this function is shown in Figure 2.5 for values of the parameters such that $(E/B) = 1$ Å. It should be noted there is a very slow decay of θ to zero with increasing b, which is a characteristic of the long-range influence of the coulomb potential.

We turn now to an investigation of the deflection angle for potentials which possess an attractive as well as a repulsive region. This is the most common type of potential that is found. Either the attractive region of the potential is a shallow

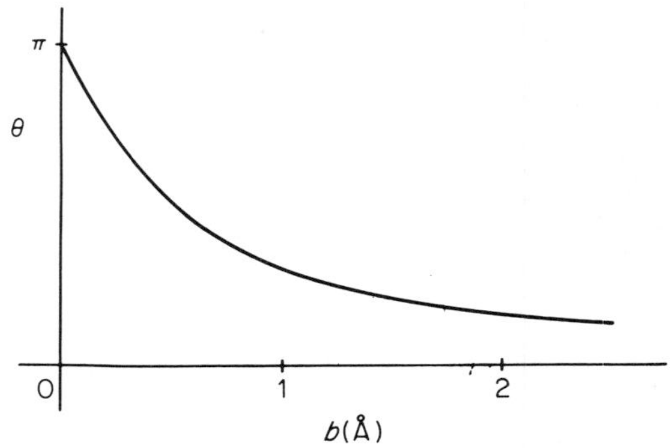

Figure 2.5 Deflection function for repulsive coulomb potential ($E/B = 1$ Å)

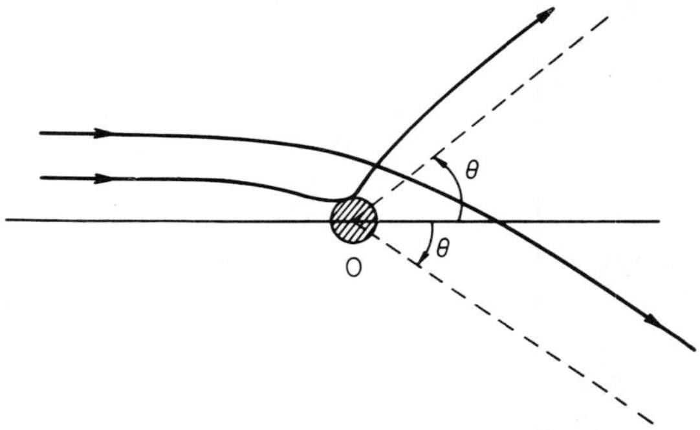

Figure 2.6 Typical trajectories for atom–atom potentials with attractive wells

well as illustrated by the van der Waals interaction of closed-shell atoms (e.g. Figure 2.2) or it is a deep well such as that associated with stable diatomic molecules.

Figure 2.6 shows two typical trajectories which arise from a potential possessing an attractive well. As the trajectory enters the attractive region the kinetic energy increases and hence the relative velocity v increases. Because the angular momentum is a constant of the motion, at a given value of ϑ, r must be decreased from its value for free motion; in effect the trajectory is pulled towards the origin of the potential. The trajectory then moves into the repulsive region of the potential, where r is increased to balance the decrease in v. This continues until the classical turning point is reached and then the reverse pattern is followed on the outward part of the trajectory.

If the trajectory is dominated by the attractive region of the potential then this will give a negative deflection angle.† This is the situation for large impact parameters. It can be seen from Figure 2.2, that at large values of b (large L) the classical turning point is greater than the value of r at which $V(r)$ switches from

†The polar angle ϑ is normally defined between 0 and π, but to preserve the continuity with respect to changes in b, θ is defined between π and $-\infty$.

being attractive to being repulsive so that only the attractive part of the potential is probed by the trajectory.

We have already shown that for any potential in which there is a repulsive region θ has the value π at $b=0$ and approaches zero as b goes to infinity. It follows that for potentials with an attractive region, θ approaches zero from below, i.e. from negative values, as b approaches infinity.

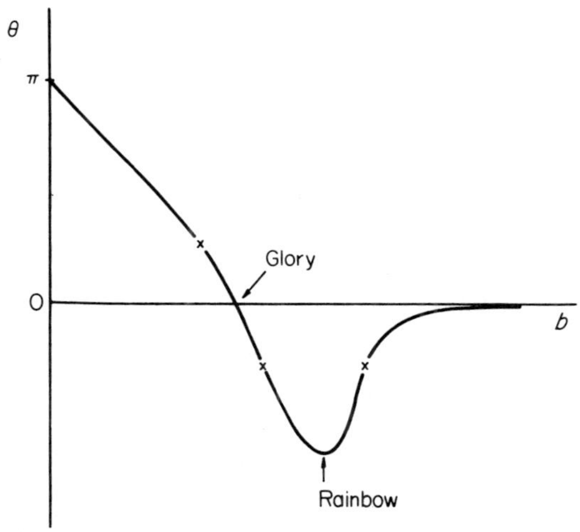

Figure 2.7 Typical deflection function for atom–atom potentials when the collision energy is large compared with the well depth. The crosses show the three impact parameters that scatter into $\theta = 50°$

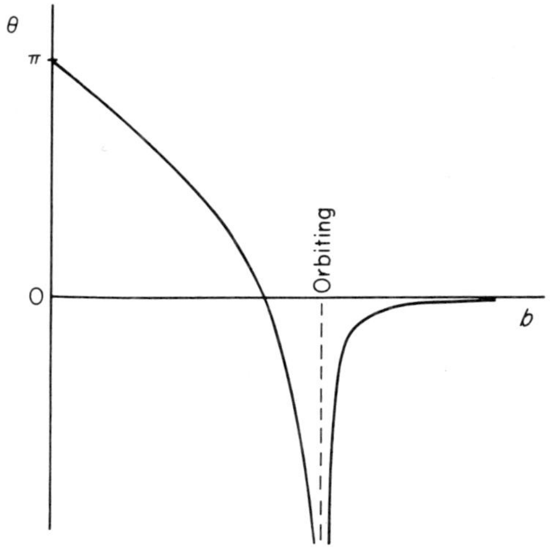

Figure 2.8 Typical deflection function for atom–atom potentials when the collision energy is small compared with the well depth

The simplest function having this behaviour is one with a single turning point, as illustrated in Figure 2.7, and this is the normal type of deflection function for atom–atom scattering. Of particular note in this figure are the points where θ is zero and where $(d\theta/db) = 0$; we shall have more to say about these in the next section.

At low collision energies there may be values of the impact parameter for which the particle orbits the target one or more times before departing. This occurs at b values for which the height of the centrifugal barrier is just less than the collision energy, for then the particle slowly enters the well of the effective potential $V_L(r)$ and stays a long time before emerging. The deflection angle for such orbiting trajectories is a rapidly varying function of b. In fact, $\theta(E, b)$ approaches $-\infty$ when the centrifugal barrier is exactly equal to the collision energy, as shown in Figure 2.8. As there is a maximum height for a centrifugal barrier (see Figure 2.2d) there will be no orbiting trajectories at high collision energies.

2.2 Collision cross-sections

The deflection angle is a mathematical quantity associated with classical trajectories and is not directly observed in a laboratory experiment. This is, firstly, because experiments do not show the result of a single collision with a specific value for the impact parameter, but rather the result of many such single collisions with all possible impact parameters. Secondly, for a central field potential we cannot distinguish between the deflection angles θ and $-\theta$ or, to include orbiting trajectories, between either of these and $\theta - 2n\pi$. The scattering is cylindrically symmetrical about the z axis of Figure 2.1.

The scattering angle Θ, which is the modulus of θ in the range 0 to π, is defined as the polar angle in a spherical polar coordinate system at which a cylindrically symmetrical beam of particles coming from $z = -\infty$ parallel to the z axis is scattered into a cylindrically symmetrical cone about the z axis; this is illustrated in Figure 2.9. From a scattering experiment we can determine the distribution of scattering angles, and this is related to a quantity called the differential cross-section $\sigma(\Theta)$, which is defined as the ratio of the scattered flux of particles (number of particles per unit time) per unit solid angle to the incident flux per unit area.

Let us now examine Figure 2.9 which has been drawn so that *all* particles lying in the cylindrical section between b and $b + db$, which has an area

$$dA = 2\pi b\, db \qquad 2.20$$

are scattered into the solid angle

$$d\Omega = 2\pi \sin\theta\, d\theta \qquad 2.21$$

For elastic scattering the velocity of particles in the incident and scattered beams are the same so that the fluxes in dA and $d\Omega$ are equal. It follows that

$$\sigma(\Theta) = \frac{\text{scattered flux}}{d\Omega} \div \frac{\text{incident flux}}{dA} = \frac{dA}{d\Omega} = \frac{2\pi b\, db}{2\pi \sin\theta\, d\theta} \qquad 2.22$$

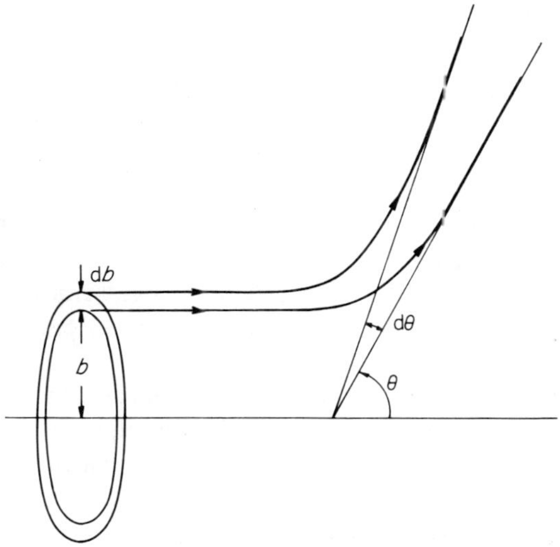

Figure 2.9 Parameters for the derivation of the cross-section for central field scattering

or

$$\sigma(\Theta) = \left| \frac{b}{\sin\theta(\mathrm{d}\theta/\mathrm{d}b)} \right| \qquad 2.23$$

where we have introduced the modulus because the cross-section is essentially positive whereas the factors $\sin\theta$ or $(\mathrm{d}\theta/\mathrm{d}b)$ can be positive or negative.

If there is more than one value of b associated with a particular scattering angle Θ then the differential cross-section at this angle is a sum of contributions such as (2.23) for each of these b values; thus

$$\sigma(\Theta) = \sum_i \left| \frac{b_i}{\sin\theta(\mathrm{d}\theta/\mathrm{d}b)_i} \right| \qquad 2.24$$

Figure 2.7 shows, for example, that there are three impact parameters that scatter into $\theta = 50°$.

The integral of the differential cross-section over the whole angular space is called the total scattering cross-section

$$\sigma = 2\pi \int_0^\pi \sigma(\Theta)\sin\Theta\,\mathrm{d}\Theta \qquad 2.25$$

As the deflection angle for a specified potential is a function of the collision energy, E, both $\sigma(\Theta)$ and σ will be functions of E although this dependence has not been shown explicitly in expressions (2.23) and (2.25).

It can be seen from Figure 2.2 that the classical turning point of the effective potential, r_C, increases as b (or L) increases. For energies above the maximum height of the centrifugal barrier (the inflection point in Figure 2.2d), r_C increases continuously as b increases; below this energy there is a discontinuity at the b

value for which orbiting occurs. In general terms we can say that the collision radius increases as b increases and in consequence the cross-section is smallest for backward scattering ($b = 0, \theta = \pi$) and increases towards forward scattering ($b \to \infty, \theta \to 0$). The cross-section is infinite for forward scattering for all 'soft' potentials, as illustrated in Figure 2.10. This figure also contains results from quantum calculations which will be described later.

It can be seen from expression (2.23) that the differential cross-section that is derived from classical mechanics has two possible singularities, one when $\sin\theta$ is zero and the other when $(\mathrm{d}\theta/\mathrm{d}b)$ is zero. These singularities will be associated experimentally with a very high intensity of scattered particles at particular angles.

When $\sin\theta = 0$ we have what is called a glory effect, and this can be seen in either forward or backward scattering.† It should, however, be noted that the backward scattering which occurs at $b = 0$ does not give a glory because the limit $(b/\sin\theta)$ is finite at $b = 0$. A backward glory is therefore only seen when orbiting trajectories occur and θ takes the values $-\pi$, -3π, etc. A forward glory is much more common as there is always one trajectory for which $\theta = 0$ if the potential has both attractive and repulsive regions, as can be seen from Figure 2.6. However, a forward glory is not easily seen as it is often masked by the intensity of non-scattered particles which also appear at $\theta = 0$.

When $(\mathrm{d}\theta/\mathrm{d}b)$ is zero the singularity that appears in the differential cross-section is called a rainbow. In Figure 2.10 this appears at an angle of approximately $\pi/2$. The simplest way of understanding this is that a broad band

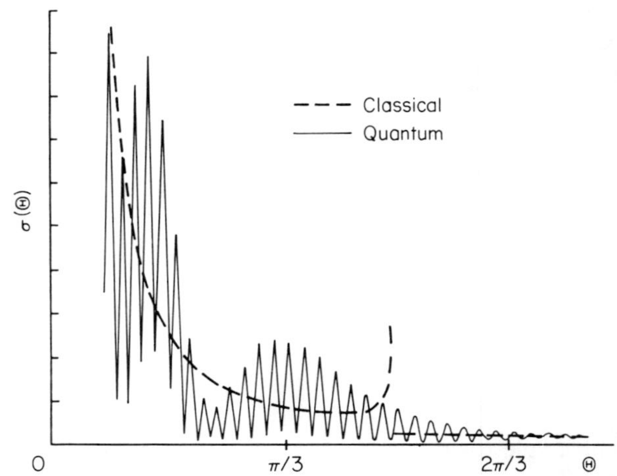

Figure 2.10 Quantum and classical cross-sections for atom–atom scattering associated with a deflection function of the type shown in Figure 2.7. The calculations were made for an Ar_2 potential and a collision energy equal to twice the well depth

†The term 'glory' is used to describe phenomena in the diffraction of light such as the halo that can appear around the moon, or the high intensity of sunlight which is backward-scattered from clouds.

δb of incident trajectories appear in a narrow band $\delta\theta$ of the scattering angle so that the intensity of scattered particles at this angle becomes very high. The analogy with the optical rainbow must be made with some caution for the optical effect is strictly a manifestation of the wave nature of light, depending as it does on the property of total internal reflection of light within a raindrop for certain angles of incidence to the surface. The similarity with the scattering rainbow extends only to the fact that at some critical angle (which for light depends on the wavelength; hence the rainbow colours) there is a very high incidence of scattered light or particles.

A rainbow singularity can only occur if there are both attractive and repulsive branches of the interatomic potential. They occur at all collision energies, but at very high energies where the effect of the repulsive branch is dominant the rainbow occurs so close to the forward scattering that it may be difficult to detect. To calculate the rainbow angle for a general interactive potential one must resort to numerical solutions of equation (2.13). Mason† was the first to show that the rainbow angle is mainly determined by the ratio of the collision energy to the depth of the potential well, D, and this can be seen in Figure 2.11 for several potentials all having the same equilibrium distance (r_e) and force constant (f_2). The rainbow angle is less sensitive to these parameters than it is to the well depth.

For the Morse and Rydberg potentials it can be shown that the rainbow angle is invariant to the product $f_2 r_e^2$. For the Lennard–Jones potential, $f_2 r_e^2 = 72D$, so that the only parameter affecting the rainbow angle is D. It has sometimes been assumed that this holds for all potentials but that is not the case.

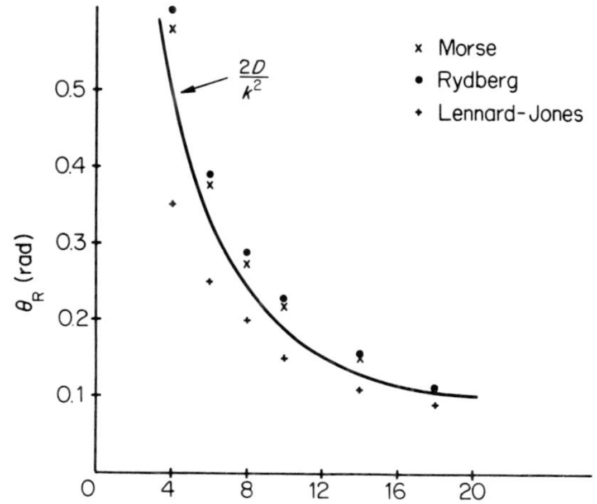

Figure 2.11 The rainbow angle as a function of the ratio of the collision energy ($k^2 = 2\mu E/\hbar^2$) to the well depth D for the Lennard–Jones potential $V = D(r^{-12} - 2r^{-6})$, and for the Morse and Rydberg functions with the same equilibrium distance ($r_e = 1$ Å) and force constant

†E. A. Mason, *J. Chem. Phys.*, **26**, 667 (1957).

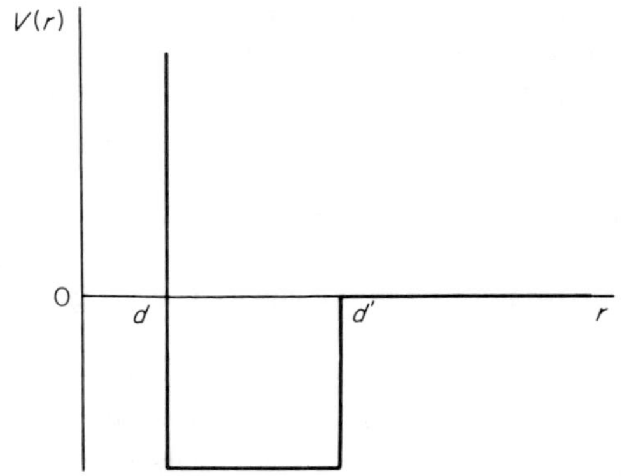

Figure 2.12 Parameters of the square well potential

It is interesting to note that the square well potential illustrated in Figure 2.12 gives negative deflection angles but does not give a rainbow. As the b value approaches the outer radius of the square well, d', the deflection angle approaches its most negative value, but it then changes discontinuously to zero at this radius; $(d\theta/db)$ is non-analytic at $b = d'$.

In the previous section we derived analytical expressions for the deflection angle for the hard sphere and coulomb potentials. From these we can obtain expressions for the differential and total cross-sections. By introducing (2.16) into (2.24) and (2.25) we obtain for the hard sphere potential

$$\sigma(\Theta) = \frac{d^2}{4}, \qquad \sigma = \pi d^2 \tag{2.26}$$

Thus the differential cross-section is isotropic and is independent of the collision energy.

The expression for the total cross-section is exactly equal to the area within which there will be a collision of two hard spheres of diameter $d/2$, so that the formal definition of cross-section is entirely consistent with our simple view of a collision in this case.

For coulomb (Rutherford) scattering the differential cross-section can be obtained from (2.19) as

$$\sigma(\Theta) = \left(\frac{B}{4E}\right)^2 \operatorname{cosec}^4\left(\frac{\Theta}{2}\right) \tag{2.27}$$

and the strong singularity at $\Theta = 0$ leads, after insertion into (2.25), to an infinite cross-section.

As the differential cross-section is infinite at rainbow and glory angles, both of which exist whenever there are attractive and repulsive branches to the potential, it might be thought that these singularities in $\sigma(\Theta)$ would give rise to infinite total cross-sections. This is not the case. In expression (2.25) the integral contains $\sin\Theta$; hence there will be no infinity from the glory. Moreover, in the vicinity of the rainbow, $\theta_R - \theta$ is proportional to $(b - b_R)^2$ and $\sigma(\Theta)$ is, from (2.24), proportional to $b_R/(\theta_R - \theta)^{1/2} \sin \theta_R$. The singularity around θ_R is therefore integrable.

However, classical mechanics gives infinite cross-sections whenever there is an attractive long-range potential. It can be shown† that the contribution to forward scattering from collisions at high impact parameters with an attractive potential r^{-s}, $s \geqslant 2$, is proportional to $\theta^{-2(s+1)/s}$, which is not integrable. Physically, any attractive potential which is not cut off at finite r will give a deflection no matter how large the impact parameter, and therefore forward scattering always gives an infinite cross-section. In quantum mechanics small deflections have no physical meaning, by the uncertainty principle, and hence total cross-sections calculated by quantum mechanics for such potentials are generally finite. The coulomb potential is the exception as this gives an infinite cross-section even in quantum mechanics. The coulomb potential falls off at large r more slowly than the centrifugal potential and, as noted in the next section, requires special treatment.

2.3 Quantum scattering by a central force

The inherent defect of the classical approach to scattering is contained in expression (2.13) which shows that a particular value of the impact parameter is associated with a particular deflection angle. The impact parameter is related to angular momentum by (2.11) and we know, by the uncertainty principle, that for systems of an atomic scale if we have no uncertainty in the angular momentum then we have complete uncertainty in the angular position, i.e. in the deflection angle. In other words, quantum mechanics implies that every value of the angular momentum (or impact parameter) should to some extent contribute to scattering at all angles. We anticipate from this that certain effects which appear in classical theory as singularities, such as rainbows, glories and orbiting resonances, will be smeared out over a range of angles or energies in quantum theory.

A second feature that is absent from the classical picture is the interference between trajectories that give rise to the same scattering angle. The classical expression for the differential cross-section (2.24) is just a sum over all trajectories that contribute to a particular angle whereas, as we have noted in Section 1.4, we should properly sum wave functions not probabilities. We shall return to a more detailed discussion of these points when we discuss the semi-classical approach to elastic scattering in the next section.

We anticipate that our readers are already familiar with the solution of one central force problem in quantum mechanics, which is the electronic energy levels of the hydrogen atom. The Schrodinger equation for this system is most easily solved by taking spherical polar coordinates as variables and separating the wave function into a product of angular functions, which are the spherical harmonic functions, and radial functions, which involve the associated Laguerre polynomials.

†F. T. Smith, *Phys. Rev.*, **150**, 79 (1966).

Formally the Schrodinger equation for central force scattering is very similar to that for the hydrogen atom, differing only in the replacement of the coulomb potential between an electron and a proton ($-e^2/r$) by some other function of r, which is the atom–atom potential. We might, therefore, expect the mathematical treatment of atomic scattering to be rather similar to that of the energy levels of the hydrogen atom, and in fact the *bound* states of a diatomic molecule can be represented, as for the hydrogen atom, as products of angular functions (the rotational wave functions) and functions of the internuclear distance r (the vibrational wave functions). The angular functions are again the spherical harmonics and for some special cases such as the harmonic oscillator and the Morse potential the radial functions are also standard analytical functions.

The difference between the central force scattering problem and the hydrogen atom spectrum lies not in the form of the Schrodinger equation but in the boundary conditions which we apply to the solutions. For bound-state problems the wave functions approach zero when r approaches infinity and this boundary condition produces quantization of the energy levels; i.e. only special energy levels are allowed. In the scattering problem the wave functions are not zero at infinity; rather we must have wave functions that represent the atoms coming together initially along straight lines and departing from one another in some angular pattern which may be quite complicated.

For the scattering problem we require wave functions in the continuum where there is no quantization of energy. There are solutions (wave functions) of the Schrodinger equation for the central force problem in the continuum which are products of spherical harmonics and radial wave functions but they do not represent any physically attainable scattering experiment. The solutions we seek are not separable functions of this kind but, as we shall see, they come into our analysis; they are used for their *mathematical convenience* rather than their physical reality.

We start by returning to expression (1.15) and noting that the wave function $\exp(ikz)$ represents a stream of particles moving with momentum $k\hbar$ in the direction of increasing z. This is called a plane wave and it describes the incident beam (in relative coordinates) in the idealized crossed beam experiment (by convention particles coming in along the z axis). After the collision has occurred and the atoms are well away from the collision region, we have particles moving with the same magnitude of the relative momentum as in the incident beam (because the scattering is elastic), but they are moving out in all directions.

If the particle were moving isotropically, that is with the same probability in all directions, then this would be represented by a wave function

$$\frac{\exp(ikr)}{r} \qquad 2.28$$

which is called a spherical wave. There is a clear similarity to the plane wave expression (1.15) but the r in the denominator is required to preserve probability within a given solid angle. To introduce the anisotropic character of the atomic scattering experiment we represent the situation after the collision by a wave function

$$\frac{f(\theta)\exp(ikr)}{r} \qquad 2.29$$

where $f(\theta)$ is called the scattering amplitude. Only one angular variable is needed for central force scattering; the wave function is independent of the polar angle ϕ because the scattering has cylindrical symmetry about the z axis.

We can now establish the boundary conditions on the wave functions for central force scattering through the asymptotic expression

$$\Psi \sim e^{ikz} + \frac{f(\theta)e^{ikr}}{r} \qquad 2.30$$

where $\sim$ indicates the limit as $r \to \infty$. What we do not know is the form of the wave function in the interaction region (where the interatomic potential is non-zero), and to find this we must solve the Schrodinger equation with this boundary condition. Although the wave function in the interaction region is not directly connected to any experimentally observable, there is no way of relating $f(\theta)$ to the interaction potential without going through this step.

The Schrodinger equation for a system of reduced mass μ acted on by a central potential $V(r)$ is

$$\left[\frac{-\hbar^2}{2\mu}\nabla^2(\mathbf{r}) + V(r)\right]\Psi(\mathbf{r}) = E\Psi(\mathbf{r}) \qquad 2.31$$

and by introducing the variables

$$k^2 = \frac{2\mu E}{\hbar^2}, \qquad U(r) = \frac{2\mu V(r)}{\hbar^2} \qquad 2.32$$

this takes the simpler form

$$[\nabla^2 + k^2 - U(r)]\Psi(\mathbf{r}) = 0 \qquad 2.33$$

A completely general solution of this equation can be obtained by taking an expansion of the wave function in spherical harmonic functions because this is a complete set of functions for the description of any analytic function of the polar angles:

$$\Psi(\mathbf{r}) = \sum_{l=0}^{\infty}\sum_{m=-l}^{l} R_{lm}(r)Y_{lm}(\theta, \phi) \qquad 2.34$$

However, for central force scattering there is no ϕ dependence in the wave function and only terms with $m = 0$ contribute to the expansion. The functions Y_{l0} are, apart from a multiplying factor, the Legendre functions $P_l(\cos\theta)$. The expansion for the required solution of (2.33) is therefore conventionally written

$$\Psi(\mathbf{r}) = \frac{1}{r}\sum_{l=0}^{\infty} A_l\psi_l(r)P_l(\cos\theta) \qquad 2.35$$

and is called the *partial wave expansion*. We are not implying that this expansion is rapidly convergent; we shall see later that one frequently needs at least 100 terms to obtain a good description of heavy atom scattering. However, it is a mathematically convenient expansion.

On substituting the expansion (2.35) into (2.33) we deduce that the radial wave functions $\psi_l(r)$ are solutions of the radial equation

$$\left[\frac{d^2}{dr^2} + k^2 - U_l(r)\right]\psi(r) = 0 \tag{2.36}$$

where

$$U_l(r) = U(r) + \frac{l(l+1)}{r^2} \tag{2.37}$$

which is called the effective potential. Notice its close similarity to the effective potential that occurs in classical theory (2.6), the difference only being the replacement of L^2 by $\hbar^2 l(l+1)$, a familiar feature in quantum mechanics, and multiplication throughout by $(2\mu/\hbar^2)$.

As (2.36) is a second-order differential equation there are two linearly independent solutions for each value of k^2 and l. However, only one of these solutions is acceptable and that is the one which is zero at $r = 0$. This is called the regular solution.

It is relatively easy to find solutions of (2.36) for any potential $U_l(r)$ by numerical integration, starting the wave function with zero amplitude at $r = 0$ (or in practice giving it a very small value at some value of r that is well inside the repulsive region of the potential) and integrating outwards to such a large value of r that $U_l(r)$ can be taken as zero. Figure 2.13b shows a typical wave function for a standard interaction potential. The wave function expands roughly exponentially from $r = 0$ to the classical turning point, where $k^2 = U_l(r)$, following which there is a series of oscillations whose wavelength gradually approaches a constant value. Indeed, we can see that when r reaches a value such that $U_l(r)$ can be neglected, the wave functions satisfy

$$\left(\frac{d^2}{dr^2} + k^2\right)\psi(r) = 0 \tag{2.38}$$

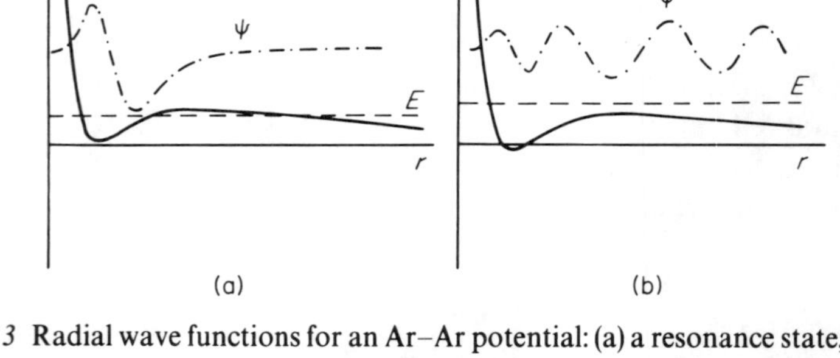

Figure 2.13 Radial wave functions for an Ar–Ar potential: (a) a resonance state, (b) a non-resonance state

which has general solutions which can be written in the form

$$\sin(kr + \eta_l) \tag{2.39}$$

where η_l is a phase which depends on k, l and the form of the potential.

We have already seen, from (2.30), that our interest is in the asymptotic form of the wave function at $r = \infty$, which is where the experimental observations are made, and there the only feature of the wave function that depends on the potential is the phase η_l. We therefore see that η_l is a very important quantity in the theory of elastic scattering. However, its value depends not only on $U(r)$ but also on the centrifugal potential, and it is therefore appropriate to take as a reference the phase that is obtained by the centrifugal potential alone, i.e. the phase for $U(r) = 0$. If $U(r)$ is zero, equation (2.36) becomes

$$\left[\frac{\mathrm{d}^2}{\mathrm{d}r^2} + k^2 - \frac{l(l+1)}{r^2}\right]\psi_l(r) = 0 \tag{2.40}$$

The solutions of this equation are well known and related to spherical Bessel functions $j_l(r)$ by the expression

$$\psi_l(r) = krj_l(r) \tag{2.41}$$

which have the asymptotic form

$$\psi_l \sim \sin\left(kr - \frac{l\pi}{2}\right) \tag{2.42}$$

We therefore define the asymptotic form of the solutions of (2.35) as

$$\psi_l(r) \sim \sin\left(kr - \frac{l\pi}{2} + \delta_l\right) \tag{2.43}$$

where

$$\delta_l = \eta_l + \frac{l\pi}{2} \tag{2.44}$$

is called the *phase shift*.

The coulomb potential is a special case for which this formula does not apply because the potential (r^{-1}) falls off more slowly than the centrifugal potential (r^{-2}) and can therefore never be neglected. Direct integration of the radial wave equation (2.36) for this case shows that the phase shift is a divergent function of r. The resulting differential cross-section for the coulomb potential is in fact identical with the classical result (2.27).

The dependence of δ_l on l is very interesting and reflects features of the potential. For example, a purely repulsive potential gives rise to a *negative* phase shift. The simplest example of this is the hard sphere potential. The solution of (2.36) in this case must be zero at the hard sphere radius d and this change of origin will give, for $l = 0$, a radial wave function

$$\sin[k(r - d)] \tag{2.45}$$

so that

$$\delta_0 = -kd \qquad 2.46$$

In fact for small values of l (when the classical turning point is governed by the hard sphere radius rather than by the centrifugal barrier) the wave function is still approximately of this form. Hence we deduce from (2.46) that

$$\delta_l = -kd + \frac{l\pi}{2} \qquad (\text{for } l \ll kd) \qquad 2.47$$

At large l the classical turning point is unaffected by the hard sphere radius so $\delta_1 \to 0$ as $l \to \infty$. This behaviour is illustrated in Figure 2.14.

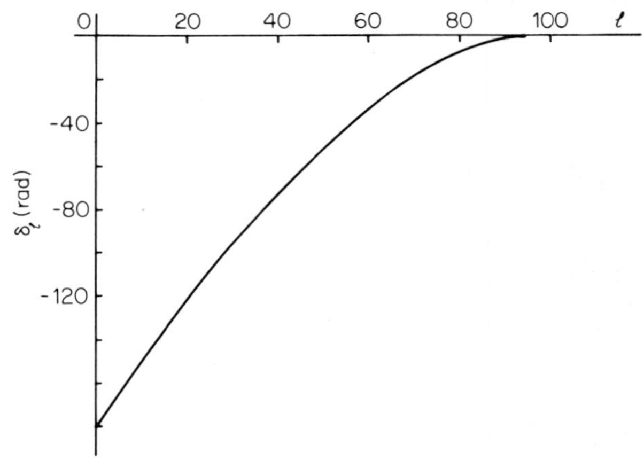

Figure 2.14 Variation of δ_l with l for a hard sphere potential

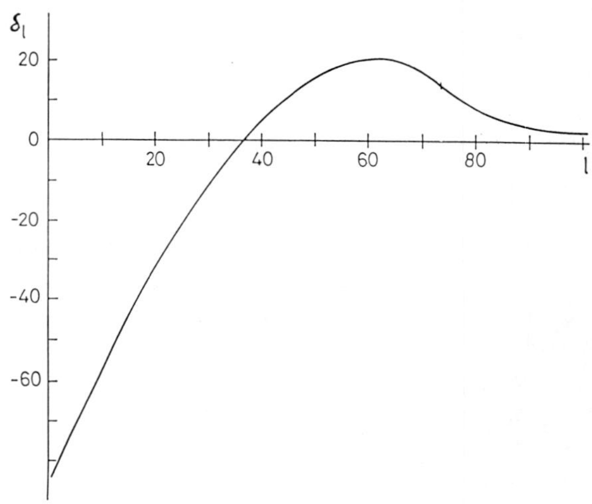

Figure 2.15 The variation in δ_1 for an Ar_2 potential. The collision energy is twice the well depth

When there is an attractive region of the potential the wave function is effectively pulled towards $r = 0$ and this produces a positive phase shift. Figure 2.15 shows the dependence of the phase shift on l for the same interatomic potential used in the classical calculation of Figure 2.7.

As an aside to the above discussion we note that there is a theorem due to Levinson which shows that the phase shift at zero energy ($k = 0$) is related to the number of bound states N_l supported by the potential $U_l(r)$ through the expression

$$\delta_l(0) = N_l \pi \qquad 2.48$$

If there is also a bound state at $k = 0$ (i.e. a state whose wave function decays to zero as $r \to \infty$) then the appropriate expression is

$$\delta_l(0) = (N_l + \tfrac{1}{2})\pi \qquad 2.49$$

Figure 2.16 shows the variation of phase shift with energy for an OH potential with $l = 50$. There are no true bound states for this effective potential as the minimum is only just below the dissociation limit. However, it can be seen that there are four energies where the phase shift changes by π over a small change of energy, i.e. it follows expression (2.48). These are the energies of metastable states or resonances in the effective potential; they are states which have a finite lifetime which is inversely proportional to the width of the resonance ($\tau = h/\Delta E$). The lifetime decreases as the energy approaches the top of the centrifugal barrier due to an increase in the probability of tunnelling through the centrifugal barrier.

To demonstrate the relationship between the scattering amplitude $f(\theta)$ (2.29) and the phase shifts δ_l we can equate the asymptotic form of (2.34) which is, from

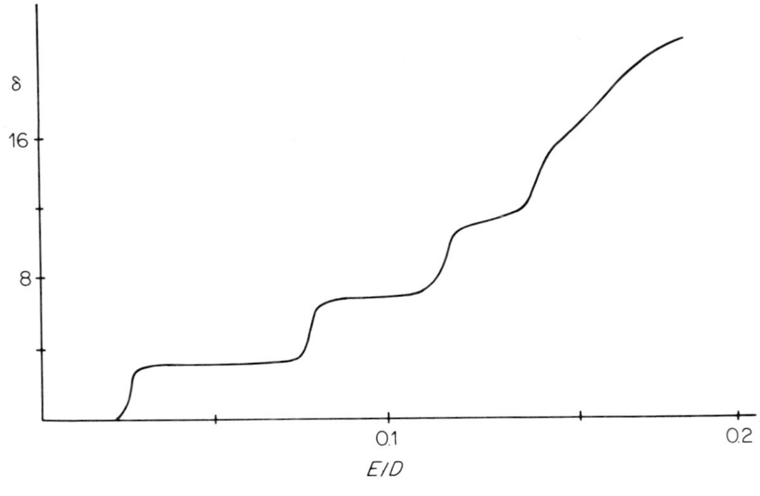

Figure 2.16 Variation of δ with energy for an OH potential with $l = 50$. The energy is in units of the dissociation energy D

(2.43),

$$\Psi(r) \sim \frac{1}{r} \sum_{l=0}^{\infty} A_l \sin\left(kr - \frac{l\pi}{2} + \delta_l\right) P_l(\cos\theta) \qquad 2.50$$

with the asymptotic form (2.30), which represents the physical situation in the crossed beam scattering experiment. To do this involves us in some standard but rather tedious algebra, and this is given in Appendix 2. The result, which is of considerable practical importance, is known as the partial wave expansion of the scattering amplitude:

$$f(\theta) = \frac{1}{2ik} \sum_{l=0}^{\infty} (2l+1)(e^{2i\delta_l} - 1) P_l(\cos\theta) \qquad 2.51$$

Of the two terms in the central factor the first, $\exp(2i\delta_l)$, is due to the interatomic potential and the second, (-1), to the contribution from the unscattered incident plane wave.

The scattering cross-section can be deduced from the amplitudes of the incident and scattered waves in the asymptotic wave function (2.30). In the incident beam the flux across unit area is proportional to

$$|e^{ikz}|^2 = 1 \qquad 2.52$$

and in the scattered beam the flux across an area of the sphere which subtends the unit solid angle at the centre is proportional to

$$r^2 \left| \frac{f(\theta) e^{ikr}}{r} \right|^2 = |f(\theta)|^2 \qquad 2.53$$

the proportionality constants being the same because the momenta of the particles are the same in the incident and scattered beams. The differential cross-section, defined earlier for classical scattering (2.22) as the ratio of the scattered flux per unit solid angle to the incident flux per unit area, is therefore

$$\sigma(\Theta) = |f(\theta)|^2 \qquad 2.54$$

On replacing $f(\theta)$ by the partial wave expansion (2.51) we therefore obtain the following expression:

$$\sigma(\theta) = \frac{1}{4k^2} \sum_{l=0}^{\infty} \sum_{l'=0}^{\infty} (2l+1)(2l'+1)(e^{-2i\delta_l} - 1)$$

$$\times (e^{2i\delta_{l'}} - 1) P_l(\cos\Theta) P_{l'}(\cos\Theta) \qquad 2.55$$

in which we note that there are interference terms between l and l' partial waves.

The expression for the total cross-section is rather simpler because on integrating over all angles, as in expression (2.24), we can make use of the orthogonality of the Legendre polynomials

$$\int_0^{\pi} P_l(\cos\theta) P_{l'}(\cos\theta) \sin\theta \, d\theta = \frac{2\delta_{ll'}}{2l+1} \qquad 2.56$$

so that only the terms with $l = l'$ persist. We therefore obtain the result

$$\sigma = \frac{\pi}{k^2} \sum_{l=0}^{\infty} (2l+1)(e^{-2i\delta_l} - 1)(e^{2i\delta_l} - 1)$$

$$= \frac{4\pi}{k^2} \sum_{l=0}^{\infty} (2l+1) \sin^2 \delta_l \qquad 2.57$$

For atomic collisions, at the collision energies normally encountered in laboratory conditions, one generally needs at least a hundred partial waves to obtain convergence of the cross-sections. An estimate of the maximum value of l which is needed can be obtained from the classical expression (2.11) by equating the impact parameter to the range of the interatomic potential; if it is greater than this there will be no deflection and hence no contribution to the cross-sections. For example, in the collision of two argon atoms, which have a reduced mass number of 20 (mass 3.3×10^{-26} kg) and a collision energy of 1 eV(1.6×10^{-19} J), we can take the maximum range of the potential to be about 10 Å. The maximum angular momentum for a deflection is then approximately 10^{-31} J s which is 10^3 in units of $\hbar$. Thus for this system one needs about a thousand partial waves to obtain convergence to the total cross-section.

Figure 2.10 shows a comparison of the quantum and classical cross-sections for Ar–Ar scattering at a collision energy equal to twice the well depth (~ 0.04 eV). Approximately 200 partial waves were needed to obtain convergence. The fine oscillations arise from the cross terms ($l \neq l'$) in (2.55) and the broad oscillation with a peak at $\theta \approx \pi/3$ is associated with the rainbow, as will be shown in Section 2.5.

With so many terms in the partial wave expansion to be considered it is not surprising that approximations have been found that simplify the problem. The simplest is to replace the sum in (2.55) by an integral. Providing that the phase shift δ_l is a smooth function of l, an accurate value of the required integrals can be obtained without having its value at each integer value of l: each interval of ten might, for example, be sufficient.

Other approximations can be made to the terms that enter the partial wave expression. For example, in the limit $l \sin\theta \gg 1$ we can replace the Legendre polynomials which appear in (2.51) by the asymptotic expression

$$P_l(\cos\theta) \approx \left(\frac{2}{\pi l \sin\theta}\right)^{1/2} \cos\left[\left(l + \frac{1}{2}\right)\theta - \frac{\pi}{4}\right] \qquad 2.58$$

We can also make use of the sum rule for these polynomials

$$\sum_{l=0}^{\infty} (2l+1) P_l(\cos\theta) = 0, \qquad \theta \neq 0 \qquad 2.59$$

to remove the contribution to $f(\theta)$ from the unscattered wave. Using both (2.51) and (2.58) the scattering amplitude can be approximated by

$$f(\theta) \approx \frac{1}{2ik} \left(\frac{2}{\pi \sin\theta}\right)^{1/2} \int_0^{\infty} \left(\frac{2l+1}{l^{1/2}}\right) e^{2i\delta_l} \cos\left[\left(l + \frac{1}{2}\right)\theta - \frac{\pi}{4}\right] dl \qquad 2.60$$

although we must remember that this will be invalid for small θ and that it will not accurately represent the contributions from low-order partial waves. By writing the cosine as the sum of two exponentials, replacing i by $e^{i\pi/2}$ and replacing $(l+\frac{1}{2})$ by l, (2.60) becomes

$$f(\theta) \approx \frac{1}{k(2\pi \sin \theta)^{1/2}} \int_0^\infty l^{1/2}[e^{iv_+(l)} + e^{iv_-(l)}]\, dl \qquad 2.61$$

where

$$v_\pm(l) = 2\delta_l - \frac{\pi}{2} \pm \left(l\theta - \frac{\pi}{4}\right) \qquad 2.62$$

We can now introduce what is called the stationary phase approximation to the integral. This is based on the observation that the regions of l that make the most important contributions to the integral are those for which $v_+(l)$ or $v_-(l)$ are stationary with respect to changes in l. The reason for this is that for all other regions the integrand oscillates rapidly with changing l and the contributions from these approximately cancel. Thus the only terms that are retained in (2.61) are those for which

$$\frac{\partial v_\pm(l)}{\partial l} = \frac{2\partial\delta_l}{\partial l} \pm \theta = 0 \qquad 2.63$$

We have seen (e.g. in Figure 2.15) that the derivative of δ_l with respect to l can be positive or negative; generally it is positive for small l and negative for large l. Thus there may be solutions of (2.63) at more than one value of l, and this is analogous to the fact that more than one classical trajectory (b value) may contribute to a particular scattering angle. However, expression (2.61) retains the phenomenon of interference between these stationary values, because the cross-sections are obtained from the square modulus of $f(\theta)$.

If l_0 is a stationary point of $v(l)$ [i.e. satisfies (2.63)] then

$$v(l) \approx v(l_0) + \tfrac{1}{2}(l - l_0)^2 v''(l_0) \qquad 2.64$$

and

$$\int_0^\infty dl \cdot l^{1/2} e^{iv(l)} \approx e^{iv(l_0)} \int_0^\infty dl \cdot l^{1/2} e^{i(l-l_0)^2 v''(l_0)/2}$$

$$\approx \left[\frac{2\pi l_0}{v''(l_0)}\right]^{1/2} \exp\left[iv(l_0) + i\frac{\pi}{4}\right] \qquad 2.65$$

For the second of these approximations it is necessary that the integrand falls rapidly to zero away from $l = l_0$ [i.e. $v''(l_0)$ is not close to zero] and that l_0 is large.†

† $$\int_0^\infty dl \cdot l^{1/2} e^{i(l-l_0)^2 v''(l_0)/2} \approx l_0^{1/2} \int_{-\infty}^\infty dl\, e^{il^2 v''(l_0)/2} = \left[\frac{2\pi l_0}{v''(l_0)}\right]^{1/2} e^{i\pi/4}$$

Substituting this result into (2.61) and taking the stationary points of either v_+ or v_- gives

$$f(\theta) \approx \sum_{l_0} \frac{l_0^{1/2} e^{i(v_\pm(l_0) + \pi/4)}}{k[v''_\pm(l_0) \sin\theta]^{1/2}} \qquad 2.66$$

the summation being over each stationary point. If two stationary points are close together (i.e. v''/l_0 is small) then a more accurate method is needed which is known as the uniform approximation. If there are many stationary points close together, as in the case of orbiting (Figure 2.8), then the classical and semi-classical models fail completely.

2.4 A semi-classical view of elastic scattering

By classical mechanics we deduce, from expression (2.24), that the particles emerging at a particular value of the scattering angle are associated with one, or at most a few, specific values of the impact parameter or angular momentum. From quantum mechanics we deduce, from expression (2.51), that all values of the angular momentum contribute to scattering at all angles. As atoms are heavy by quantum mechanical standards (compared with electrons, for example) we expect the classical picture to be not totally misleading. It should, therefore, be possible to show that even in quantum mechanics the scattering at a particular angle will often be dominated by one, or a few, or a small interval of the angular momentum.

The mathematical apparatus for proving this is provided by an approximation to quantum mechanics called semi-classical theory which is based on the fact that Planck's constant, h, is very small. Behind this approximation is the observation that all quantum systems must show classical behaviour when the masses of the particles are large, and since the ratio h^2/mass always appears in such an analysis this is equivalent to examining the limit $h \to 0$. The manifestation of this semi-classical limit in quantum theory appears when the wave functions oscillate rapidly over a region in which the potential is approximately constant. For example, an iodine atom with a kinetic energy of 1 eV has a wave function with wave length 0.1 Å, and this is very small compared with the range, several angstroms, over which an iodine atom interacts with other atoms. Iodine atom collisions at such an energy should therefore be adequately described by semi-classical mechanics.

It is also possible to approach semi-classical theory from classical mechanics, using the concept of the classical trajectory and introducing new postulates to take account of the interference between trajectories. These postulates are suggested by the equations of wave motion and, in particular, the fact that the rays which are perpendicular to a wave front obey the classical equations of motion. For our immediate objectives the development of semi-classical mechanics from quantum mechanics is the more straightforward procedure.

The semi-classical approximation was developed very early (1926) from the full equations of quantum mechanics by Wentzel, Kramers and Brillouin, but as the mathematics follows an earlier analysis (1923) of the equations of wave motion by Jeffreys, the approach is often referred to as the JWKB method.

By writing the solution of the one-dimensional Schrodinger equation in the form

$$\Psi = e^{iS(r)/\hbar} \qquad 2.67$$

and expanding $S(r)$ as a power series in $\hbar$, we find that to first order in $\hbar$, the wave function is

$$\Psi_{\pm} = [k(r)]^{-1/2} \exp\left[\pm i \int k(r)\,dr \right] \qquad 2.68$$

where

$$k(r) = \frac{p(r)}{\hbar} = \frac{[2\mu(E - V)]^{1/2}}{\hbar} \qquad 2.69$$

This result is proved in Appendix 1. The most general JWKB wave function is a linear combination of Ψ_+ and Ψ_- with coefficients determined by other factors such as the boundary conditions.

The integral that appears in (2.68) is identical to the action integral (1.29) in appropriate units. The factor $[k(r)]^{-1/2}$ provides for the conservation of flux in the particle beam.

The JWKB solution (2.68) is valid both in the classically accessible region ($E > V$, k real) and in the classically inaccessible region ($E < V$, k imaginary), but it is clearly invalid at the classical turning point ($E = V$, $k = 0$). To obtain a semi-classical approximation to the wave function which goes from a classically forbidden to a classically allowed region we therefore need to make a connection between the two, for without that we cannot determine the appropriate combination of Ψ_+ and Ψ_- in the asymptotic region which is required to calculate the phase shift and the scattering cross-section.

The way forward is to assume that in the region of the classical turning point the potential $V(r)$ has a simple form for which the solutions of the Schrodinger equation are known. There may be one or several classical turning points, depending on the energy, the angular momentum and the nature of the potential, and if there is more than one then they may be close together (in r) or far apart. Each of these situations requires individual attention as the appropriate connection formulae are different for each case.

The simplest and most common situation is for a single classical turning point (at $r = a$, say) where $r < a$ is classically forbidden and $r > a$ is classically allowed; this is called a left-hand turning point. The connection in this case is made by taking $V(r)$ to be linear in the region of $r = a$, say:

$$V(r) = E - \gamma(r - a) \qquad 2.70$$

The Schrodinger equation for this potential is

$$-\frac{\hbar^2}{2\mu}\frac{d^2\Psi(r)}{dr^2} - \gamma(r - a)\Psi(r) = 0 \qquad 2.71$$

Now the solutions of the differential equation

$$\omega(z)'' - z\omega(z) = 0 \tag{2.72}$$

are well known and called Airy functions. They occur as pairs of independent solutions, Ai(z) and Bi(z), but only the first of these is finite for all real z and hence can be a physically acceptable solution of the Schrodinger equation. We transform (2.71) to (2.72) by the substitution

$$z = -\left(\frac{2\mu\gamma}{\hbar^2}\right)^{1/3}(r-a) \tag{2.73}$$

so that in the region of the crossing point the wave function is

$$\Psi(r) = \mathrm{Ai}(z) = \mathrm{Ai}\left[-\left(\frac{2\mu\gamma}{\hbar^2}\right)^{1/3}(r-a)\right] \tag{2.74}$$

The function Ai(z) is illustrated in Figure 2.17.† It decays exponentially to zero for $z>0$ [which from (2.73) implies $r<a$] and oscillates to infinity for $z<0$ ($r>a$). Thus (2.74) has the expected behaviour of a solution of the Schrodinger equation for a potential giving rise to a left-hand turning point.

The quantity of interest in the scattering problem is the asymptotic form of the wave function at $r=\infty$. The asymptotic limit of the Airy function as z approaches $-\infty$ is known to be

$$\mathrm{Ai}(z\to-\infty) = \pi^{-1/2}(-z)^{-1/4}\sin\left[\frac{2}{3}(-z)^{3/2}+\frac{\pi}{4}\right] \tag{2.75}$$

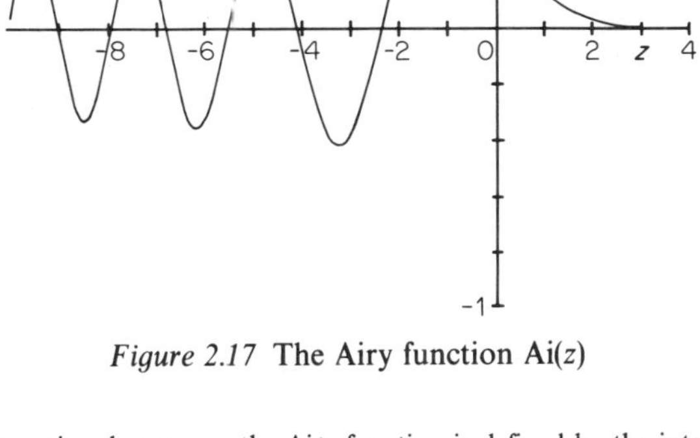

Figure 2.17 The Airy function Ai(z)

†Note that for computational purposes the Airy function is defined by the integral

$$\mathrm{Ai}(z) = \frac{1}{\pi}\int_0^\infty \cos\left(\frac{u^3}{3}+zu\right)du\,.$$

We can replace $-z$ by $r-a$ from (2.73) and rewrite $(r-a)^{3/2}$ as an integral of $(r-a)^{1/2}$. Using (2.69) and (2.70) we see that $(r-a)^{1/2}$ is proportional to $k(r)$ and on making the appropriate substitutions we obtain the asymptotic form (for $r \gg a$)

$$\psi(r) \sim [k(r)]^{-1/2} \sin\left[\int_a^r k(r)\,\mathrm{d}r + \frac{\pi}{4}\right] \qquad 2.76$$

which is seen to be a particular combination of the JWKB solutions (2.68).

To calculate the JWKB phase shift we therefore need to evaluate the action integral from the classical turning point ($r = a$) to some large value of r and add $\pi/4$, which can be thought of as the contribution to the phase from the passage through the classical barrier.

If we want the phase shift for the l partial wave we have to determine the wave vector $k(r)$ for the appropriate effective potential. From (2.69), (2.32) and (2.37) we have

$$k_l(r)^2 = \frac{2\mu(E-V)}{\hbar^2} = k^2 - U_l(r) = k^2 - U(r) - \frac{l(l+1)}{r^2} \qquad 2.77$$

However, if we use this expression to evaluate the integral in (2.76) for the case when $U(r) = 0$ we do not get the required answer (2.42), namely a phase of $-l\pi/2$. To put this right we have to make what is called the Langer correction to (2.77) when calculating the semi-classical phase shift, replacing the centrifugal potential by $(l+\frac{1}{2})^2/r^2$. The expression for $k(r)$ which has to be used in (2.76) is therefore

$$k_l(r) = \left[k^2 - U(r) - \frac{(l+\frac{1}{2})^2}{r^2}\right]^{1/2} \qquad 2.78$$

We obtain a semi-classical expression for the phase shift by equating the arguments of (2.43) and (2.76):

$$kr - \frac{l\pi}{2} + \delta_l = \int_{a_l}^r k_l(r)\,\mathrm{d}r + \frac{\pi}{4} \qquad 2.79$$

or

$$\delta_l = \int_{a_l}^{\infty} [k_l(r) - k]\,\mathrm{d}r - ka_l + \frac{(l+\frac{1}{2})\pi}{2} \qquad 2.80$$

where we have now taken the upper limit of the integral as infinity, because the phase shift is evaluated at large r where $k_l(r) = k$.

The JWKB phase shifts can be used to calculate the scattering amplitude $f(\theta)$ in the stationary phase approximation by using (2.61) and (2.63). Differentiating (2.80),† remembering that a_l is the value of r for which k_l is zero, we

†We use the Leibniz rule

$$\frac{\mathrm{d}}{\mathrm{d}x}\int_{u(x)}^{v(x)} f(x,t)\,\mathrm{d}t = f(x,v)\frac{\mathrm{d}v}{\mathrm{d}x} - f(x,u)\frac{\mathrm{d}u}{\mathrm{d}x} + \int_{u(x)}^{v(x)} \frac{\partial f(x,t)}{\partial x}\,\mathrm{d}t$$

have

$$2\frac{\partial \delta_l}{\partial l} = \pi - 2\int_{a_l}^{\infty} \frac{(l+\frac{1}{2})\,\mathrm{d}r}{r^2 k_l(r)} \qquad 2.81$$

From (2.63) this quantity is equal to the deflection angle θ, and if we write, from (2.77),

$$k_l = \frac{[2\mu(E - V_{\mathrm{L}})]^{1/2}}{\hbar} \qquad 2.82$$

and put $L = \hbar(l+\frac{1}{2})$, we reproduce the classical expression for this quantity (2.10). In other words, within the JWKB approximation the points of stationary phase for a given deflection angle can be identified with the classical trajectory whose impact parameter is

$$b = \frac{L}{\mu v} = \frac{l+\frac{1}{2}}{k} \qquad 2.83$$

The semi-classical result with stationary phase is identical to the classical result except for the fact that the scattering amplitude $f(\theta)$ is given from (2.66) as a sum of stationary phase terms and hence introduces interference in the scattering cross-section $|f(\theta)|^2$.

2.5 Comparison of classical, semi-classical and quantum cross-sections

In this section we compare the differential and total cross-sections in the classical, semi-classical and quantum models for Ar–Ar scattering at a collision energy of 24 meV using the same potential for which the wave functions of Figure 2.13 were obtained. This potential has a well depth of 12 meV.

The JWKB phase shifts and classical deflection angle are shown as a function of l (or b) in Figure 2.18. The rainbow angle occurs at the minimum of the deflection angle which is a point of inflection in δ_l. The glory angle occurs at the maximum of the phase shift or the zero in the deflection angle.

For scattering angles between π and $|\theta_{\mathrm{R}}|$ there is only one stationary phase point. From Figure 2.18 it can be seen that δ_l' is positive in this region; the appropriate solution of (2.63) will be $v_-(l_0)$ and hence the differential cross-section in the JWKB stationary phase approximation is, from (2.66),

$$\sigma(\Theta) = |f(\theta)|^2 = \frac{l_0}{k^2|v_-''(l_0)|\sin\theta} = \frac{l_0}{k^2|2\delta''(l_0)|\sin\theta} \qquad 2.84$$

As, from (2.81), $2\delta_l''$ is equal to $(\mathrm{d}\theta/\mathrm{d}l)$, this is identical to the classical cross-section (2.23) if we replace l_0 by bk where b is the impact parameter for the classical scattering angle θ.

The infinite cross-section predicted from this formula for the rainbow angle, where $(\mathrm{d}\theta/\mathrm{d}l) = k^{-1}(\mathrm{d}\theta/\mathrm{d}b) = 0$, arises from the breakdown of the stationary phase approximation, as expressed by (2.64), when two stationary points

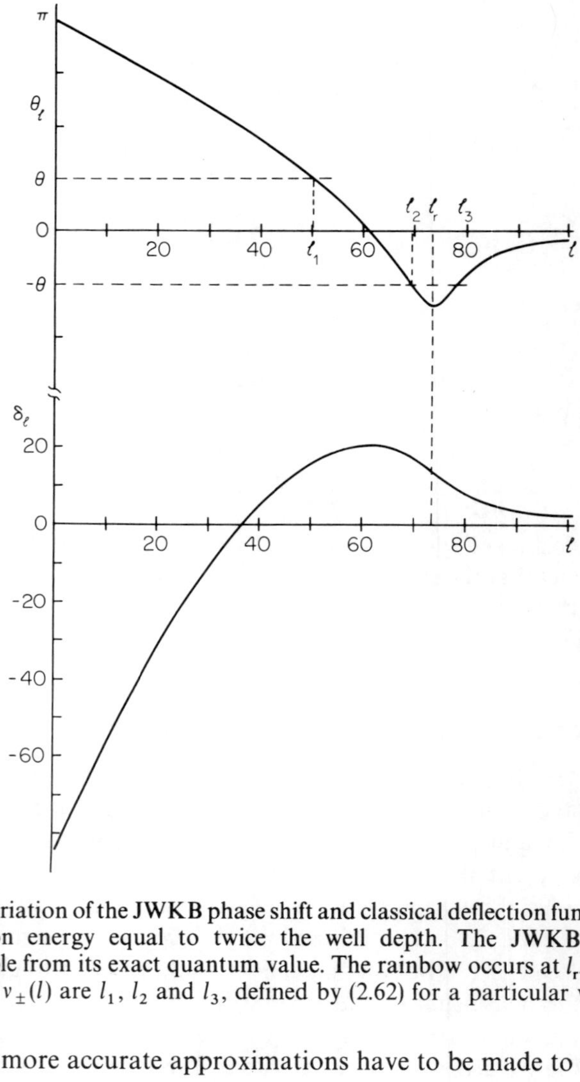

Figure 2.18 Variation of the **JWKB** phase shift and classical deflection function for Ar–Ar with a collision energy equal to twice the well depth. The **JWKB** phase shift is indistinguishable from its exact quantum value. The rainbow occurs at l_r. The stationary points of $v_{\pm}(l)$ are l_1, l_2 and l_3, defined by (2.62) for a particular value of θ

coincide, and more accurate approximations have to be made to the integral in (2.61).

For angles smaller than the rainbow angle there are generally three stationary points which contribute to the scattering amplitude. If these stationary points are at l_1, l_2 and l_3, as illustrated in Figure 2.18, then (2.60) becomes

$$f(\theta)=\frac{e^{i\pi/4}}{k(\sin\theta)^{1/2}}\left\{\left[\frac{l_1}{v''_-(l_1)}\right]^{1/2}e^{iv_-(l_1)}+\left[\frac{l_2}{v''_+(l_2)}\right]^{1/2}e^{iv_+(l_2)}+\left[\frac{l_3}{v''_+(l_3)}\right]^{1/2}e^{iv_+(l_3)}\right\} \tag{2.85}$$

The oscillations in $\sigma(\theta)$ depend mainly on the differences between the three exponents $v_{\pm}(l_i)$ that occur in (2.85). Ignoring any differences in l_i and $v''(l_i)$, there will be terms in $|f(\theta)|^2$ varying as $\cos[v_{\pm}(l_i)-v_{\pm}(l_j)]$. From (2.62) we have, for the three stationary points

$$v_-(l_1) = 2\delta(l_1) - l_1\theta - \frac{\pi}{4}$$

$$v_+(l_2) = 2\delta(l_2) + l_2\theta - \frac{3\pi}{4} \qquad 2.86$$

$$v_+(l_3) = 2\delta(l_3) + l_3\theta - \frac{3\pi}{4}$$

Close to the rainbow angle $l_2 \simeq l_3$ and $\delta(l_2) \simeq \delta(l_3)$; close to the glory angle ($\theta = 0$) $l_1 \simeq l_2$ and $\delta(l_1) \simeq \delta(l_2)$.

Figure 2.19 shows the semi-classical and the full quantum cross-section for Ar_2. The high-frequency oscillations, which are called diffraction oscillations, are due mainly to the differences between $v_-(l_1)$ and $v_+(l_2)$. The low-frequency oscillations, called supernumerary rainbows, depend mainly on the difference between $v_+(l_2)$ and $v_+(l_3)$.

There is good agreement between the semi-classical and quantum cross-sections except at the rainbow angle. We remind readers that at the rainbow angle the stationary phase points l_2 and l_3 coincide and expression (2.66) must be replaced by a better approximation known as the uniform approximation. Figure 2.10 also shows the classical cross section deduced from expression (2.24) or (2.84), and this is seen to follow the semi-classical profile in broad outline but without the oscillations. For heavy atoms or high energies the quantum oscillations become close together and they may not be resolved in the crossed beam experiment. In that case the experimental cross-section approaches the classical limit.

The position of the rainbow angle is sensitive to the collision energy. At high energies it moves towards small scattering angles and at low energies it approaches π. If the energy is sufficiently small that there is a value of l for which the centrifugal barrier just matches the energy, then classically there will be

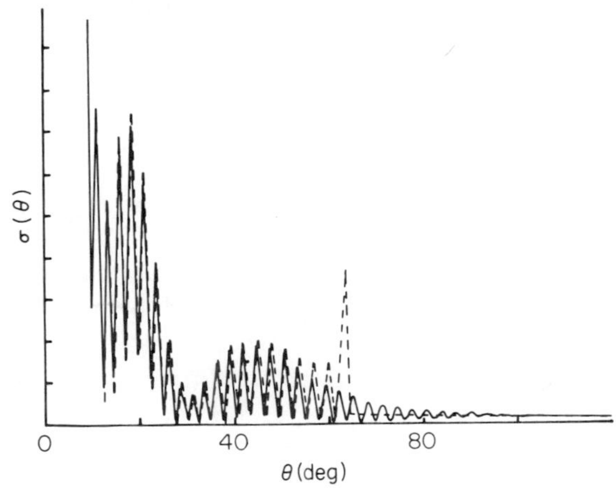

Figure 2.19 Comparison of the semi-classical (dashed) and quantum cross-section for Ar_2. Parameters are those used for Figure 2.18

orbiting trajectories, as illustrated in Figure 2.8, and the phase shift is a discontinuous function of l. The scattering amplitude is, from (2.66), the sum of an infinite number of terms, but at this point the stationary phase model breaks down because the stationary oints are not isolated.

In orbiting situations the differential cross-section is a rapidly oscillating function over the entire range of scattering angle, and it has in addition two prominent features. The first is a backward glory which is a strong peak at $\theta = \pi$ which is associated with classical trajectories that pass once, or more, completely around the target. The second is the occurrence of resonances at discrete energies due to the matching of the energy with a standing wave in the effective potential. By the quantum mechanical tunnel effect the colliding atoms can become trapped within the centrifugal barrier, and when this occurs there is a rapid change of the cross-section over a small range of energy. Resonances can clearly not be explained without quantum mechanical tunnelling.

We cannot use expression (2.66) for small scattering angles ($\theta \approx 0$) because in its derivation we used (2.58), which is only valid for large θ. For small θ we can replace $P_l(\cos\theta)$ by

$$P_l(\cos\theta) \approx 1 - \frac{l^2\theta^2}{4} + \cdots \qquad 2.87$$

in which case the scattering amplitude is approximately

$$f(\theta) \approx \frac{1}{ik}\int_0^\infty l(e^{2i\delta_l} - 1)\left(1 - \frac{l^2\theta^2}{4}\right) dl \qquad 2.88$$

After partial integration this becomes

$$f(\theta) \approx \frac{2}{k}\int_0^\infty e^{2i\delta_l}\frac{d\delta_l}{dl}\left(\frac{l^2}{2} - \frac{l^4\theta^2}{16}\right) dl \qquad 2.89$$

The stationary phase approximation can also be made to this integral by finding the points where the integrand is slowly varying. One of these is the forward glory $(d\delta_l/dl) = 0$ and the other occurs at large values of l where $\delta \approx 0$. Expressions for $f(\theta)$ can be derived for both of these and details can be found elsewhere.† The important point is that the forward scattering intensity is finite in contrast with the classical result.

2.6 The inversion problem

The treatment we have given in this chapter of elastic scattering takes us from a specified interatomic potential to calculated differential and total cross-sections. However, the principal objective in studying collisions is to determine potentials from measured cross-sections. To move in this direction we come up against what is called the inversion problem. The problem is not confined to scattering theory but occurs for many measurements in physics. In spectroscopy, for example, one measures the absorption or emission of light as a function of frequency with the aim, say, of determining molecular force constants.

†See M. S. Child, *Molecular Collision Theory*, Chap. 5.3, Academic Press, 1974.

It seems obvious that if one has experimentally measured differential cross-sections for atom–atom collisions over a wide range of collision energy then these would uniquely determine a potential. However, a general proof of this has not been found in quantum mechanics and it is only within certain approximations that the inversion problem has been solved. In the classical limit the expression for the deflection function (2.13) can be inverted provided the energy is above the orbiting limit.† The first problem is to derive the deflection function from the differential cross-section. If there is only one classical path for each scattering angle then (2.23) can be used to determine $\theta(b)$. Firstly, we have

$$b\,\mathrm{d}b = \sigma(\Theta)\sin\theta\,\mathrm{d}\theta \qquad 2.90$$

which can be integrated to give

$$\tfrac{1}{2}b^2 = \int_{\theta}^{\pi} \sigma(\Theta)\sin\theta\,\mathrm{d}\theta \qquad 2.91$$

and if there is a unique b for every θ, this can be inverted to give $\theta(b)$.

Except for repulsive potentials, or attractive potentials at high collision energies when forward scattering is ignored, there is always more than one classical path for a given scattering angle so that θ is a multivalued (normally three-valued) function of b at angles less than the rainbow angle. There is no unique direct method for the inversion under these circumstances. In quantum mechanics we encounter a similar problem, which is to determine the scattering amplitude $f(\theta)$ from the cross-section; it is $f(\theta)$ that is related to phase shifts and hence to the potential. However, taking the square root of $\sigma(\Theta)$ gives no information about the phase of $f(\theta)$.

The practical methods of determining potentials from cross-sections all involve some assumptions about the analytic form of the potential. This is not in fact a serious problem because there are many other sources that give information about potentials. For example, from spectroscopic studies of the bound states of potentials there is an accurate method based on semi-classical mechanics (the Rydberg–Klein–Reese method) by which one can deduce the function $V(r)$ within the potential well. Moreover, from quantum mechanics we know that at large interatomic distances potentials vary as r^{-6} for closed-shell atoms (more generally as r^{-n} where n is an integer). Forward scattering is predominantly determined by the tail (long-range form) of the potential. From (2.51) and (2.57) one can show that

$$\sigma = \frac{4\pi}{k}\,\mathrm{Im}\,f(0) \qquad 2.92$$

which is known as the optical theorem, so that the tail of the potential makes an important contribution to the total cross-section.

In summary, the lack of a direct inversion technique has not prevented us from combining scattering data with other information to obtain accurate interatomic potentials.

†O. B. Firsov, *J. Exp. Theory. Phys.*, **24**, 279 (1953).

Chapter 3

Inelastic collisions

3.1 Introduction

In an inelastic collision there is an exchange of energy between the translational motion of the colliding species and their internal states, but there is no exchange or transfer of atoms or transfer of charge. There are principally three types of internal state that are of interest in scattering theory, rotational, vibrational and electronic, and the energies required to change these states by one quantum are typically 10^{-2}, 10^{-1} and 1 eV respectively.

There are many situations in which electronic quanta are much smaller than 1 eV, notably for open-shell systems in which spin–orbit coupling removes the degeneracy of states, but in general one can say that electronically inelastic collisions require high beam energies, and for beams of neutral atoms or molecules these are not easy to attain. To have a significant number of molecules with 1 eV of translational energy in an equilibrium gas one requires a temperature of about 10^4 K so that very high temperatures would be required to produce a 1 eV effusive beam. High-energy ion beams are much easier to obtain because ions can be accelerated by electrostatic fields. We will examine electronic inelastic processes in Chapter 6.

Rotational and vibrational inelastic processes always go together except in very low energy beams where there is insufficient energy to span the vibrational energy spacing. The separation of internal nuclear motion into vibration and rotation is in any case an approximation which ignores Coriolis and centrifugal forces. To study rotational processes alone we can examine the rigid rotor model which has provided considerable insight into the relationship between rotational energy transfer and the features of the intermolecular potential. Vibrational inelastic processes alone can only be studied in much cruder models, e.g. in the collinear collision of an atom and a linear molecule.

In many experimental situations there are a large number of ro–vibrational levels energetically accessible from the translational energies of the colliding species. This is particularly true if one or both species is a polyatomic molecule because the rotational states are often very close together. Even if the discrete rotational states are not resolved in an experiment the average energy transferred in the collision is an important quantity for our understanding of how gases reach or depart from equilibrium populations of rotational levels. In these circumstances a strictly classical view of the internal molecular motion is likely to be sufficient to provide the information required. Indeed, the quantum treatment of rotational inelastic processes is, as we shall see, bedevilled by the problem of the large number of states that may be involved.

For the interpretation of inelastic scattering we have, as for elastic scattering, the tools of classical, semi-classical and quantum mechanics. For the most complicated inelastic processes, involving both vibrational and rotational energy transfer between molecules, exact quantum mechanical calculations are seldom feasible. For this reason it is important to make a comparison between the results obtained from the three types of mechanics on simpler systems; if the classical results, say, are valid for these, then they should also be valid for more complicated systems. We shall also see that several approximate quantum mechanical treatments can be applied to inelastic scattering, and these can often be used when exact calculations are not feasible. Again, these must be justified by comparing their performance with the exact quantum mechanical results for the simpler systems.

Our objective in this chapter will therefore be to both introduce the underlying principles of inelastic scattering and to make comparisons between the available theoretical methods for a very simple model, which is the collinear collision between an atom and a diatomic molecule. In this model energy can be transferred between the relative translational motion of the colliding species and the vibrational motion of the diatomic molecule, the so-called T–V exchange. The model does not of course represent a typical scattering process because there are no angular variables in the dynamics and therefore no scattering angle or scattering cross-section. We can, however, examine transition probabilities between energy states and deduce an average energy transfer for the collision. The relevance to real systems is that the largest T–V transfers are expected to occur when atoms collide in bond–collinear configurations.

3.2 The classical treatment of atom–diatomic molecule collinear collisions

The collinear collision of an atom and a diatomic molecule was studied by Jackson and Mott, as early as 1932, as a model for the exchange of energy in the collision between an atom and a solid surface. A landmark paper in 1966 by Secrest and Johnson provided the first exact quantum mechanical results for inelastic collisions and allowed comparison with earlier classical, semi-classical and approximate quantum mechanical results. The Secrest–Johnson treatment will be examined in Section 3.5.

Vibrationally inelastic processes are usually dominated by the short-range repulsive part of the intermolecular potential. Only strong long-range terms, as occur, for example, for ion–ion collisions, are likely to be important. The simplest model for repulsive interactions is the hard sphere collision. This was examined in Section 2.1 for elastic collisions between atoms and we will now apply it to

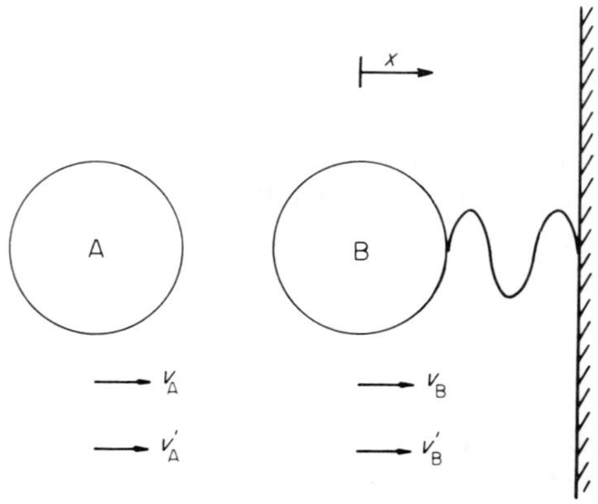

Figure 3.1 Parameters for the classical inelastic collinear collision of hard spheres

atom–molecule collision. The important feature of the hard sphere potential is that energy is transferred impulsively between the pair of atoms in collision, and this energy is later distributed within the molecule. More realistically there is some interaction between several atoms when molecules collide, and all contribute to the exchange of energy.

For simplicity we examine the hard collision between an atom A and another B which is attached to a fixed point by a harmonic spring of force constant k. This is a special case of the A + BC collision when the mass of C is infinite. There is no new physics in taking the mass of C to be finite, only the introduction of some rather tedious algebra which mainly leads to the replacement of m_A and m_B by appropriate reduced masses.[†]

Figure 3.1 defines the relevant parameters. The kinetic energy of A is

$$E_A = \tfrac{1}{2} m_A v_A^2 \qquad 3.1$$

B vibrates about its equilibrium position with a displacement that varies with time as

$$x(t) = A \sin(\omega t + \phi) \qquad 3.2$$

where A is the amplitude of the vibration, ω is the oscillator frequency given by

$$\omega = \left(\frac{k}{m_B}\right)^{1/2} \qquad 3.3$$

and ϕ is a phase. The velocity of B is the time derivative of (3.2):

$$v_B = A\omega \cos(\omega t + \phi) \qquad 3.4$$

and the oscillator energy is

$$E_B = \frac{m_B A^2 \omega^2}{2} \qquad 3.5$$

[†]See S. W. Benson, G. C. Berend and J. C. Wu, *J. Chem. Phys.*, **38**, 25 (1963).

After the collision the velocities of A and B are v'_A and v'_B respectively, and they are connected to v_A and v_B by equations based on the conservation of energy

$$m_A(v_A^2 - v_A'^2) = m_B(v_B'^2 - v_B^2) \tag{3.6}$$

and momentum

$$m_A(v_A - v'_A) = m_B(v'_B - v_B) \tag{3.7}$$

Dividing (3.6) by (3.7) and eliminating v'_A gives

$$v'_B = \frac{(m_B - m_A)v_B + 2m_A v_A}{m_A + m_B} \tag{3.8}$$

The change in energy of B on collision is entirely in its kinetic energy

$$\Delta E_B = \frac{m_B(v_B'^2 - v_B^2)}{2} \tag{3.9}$$

and replacing v'_B by (3.8) leads to

$$\Delta E_B = \frac{2m_A m_B(v_A - v_B)(m_A v_A + m_3 v_B)}{(m_A + m_B)^2} \tag{3.10}$$

If we now take the collision to occur at time $t = 0$, the outcome depends on the phase ϕ. Using expressions (3.1), (3.4) and (3.5) leads to the following result for the ratio of the energy transferred to the energy of the incoming atom:

$$\frac{\Delta E_B}{E_A} = \frac{4m_A m_B}{(m_A + m_B)^2}\left[1 + \frac{m_B - m_A}{(m_A m_B)^{1/2}}\left(\frac{E_B}{E_A}\right)^{1/2}\cos\phi - \left(\frac{E_B}{E_A}\right)\cos^2\phi\right] \tag{3.11}$$

Expression (3.11) can be positive (energy passing from A to B) or negative (energy passing from B to A) depending on the phase.

The maxima and minima points in the graph of $(\Delta E_B/E_A)$ against ϕ can be deduced by differentiating (3.11) and equating the result to zero. This leads to the expression

$$\left[2\left(\frac{E_B}{E_A}\right)\cos\phi - \frac{m_B - m_A}{(m_A m_B)^{1/2}}\left(\frac{E_B}{E_A}\right)^{1/2}\right]\sin\phi = 0 \tag{3.12}$$

which has solutions $\phi = 0$ and π, and

$$\phi = \cos^{-1}\left[\frac{m_B - m_A}{2(m_A m_B)^{1/2}}\left(\frac{E_A}{E_B}\right)^{1/2}\right] \tag{3.13}$$

The first two correspond to the collision occurring as the oscillator goes through its equilibrium position. At this position the relative velocity of A to B is either at its maximum or its minimum with corresponding extrema in the energy transfer. The turning points (3.13) only exist if the absolute value of the argument of $\cos^{-1}$ is less than one, which is generally only true if the ratio of the collision energy (E_A) to the oscillator energy (E_B) is not large. In addition, $v_A - v_B$ must be positive at the moment of impact (otherwise A would not hit B), and this is satisfied in the whole range of ϕ only if v_A is greater than the maximum velocity of B, which from (3.4) is $A\omega$. Expressing this inequality through the energies of A and B, (3.1)

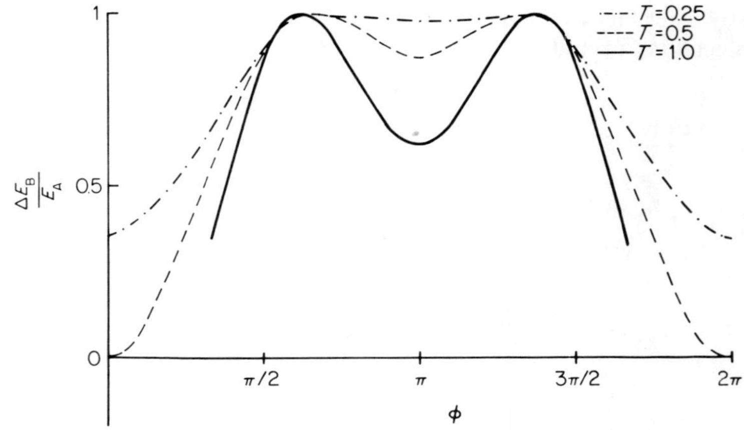

Figure 3.2 Energy transfer as a function of the oscillator phase for the collinear collision of hard spheres with mass ratio $(m_A/m_B) = 2$, and three values of T, the ratio of the oscillator energy to the kinetic energy of A

and (3.5), leads to the condition

$$\frac{E_A}{E_B} > \frac{m_A}{m_B} \tag{3.14}$$

and if this is not satisfied there will be regions of ϕ in which no collision occurs.

When the solutions (3.13) exist they represent a situation in which the whole of E_A is transferred to the oscillator so that A is stationary relative to the vibrating atom B. In such a state A will receive a further impact from B, a complication we will not pursue.

Figure 3.2 shows the energy transfer for a hard collision of two atoms with the mass ratio $(m_A/m_B) = 2$ and three values for the ratio of the oscillator energy to the kinetic energy of A. It can be seen that for the highest value of this ratio there are regions for which no collision occurs.

3.3 Numerical integration of trajectories, and action-angle variables

For the collision of an atom and a diatomic molecule using realistic potentials the equations of motion must be solved numerically, even for the model of a collinear collision. The starting point will usually be Hamilton's equations

$$\frac{dq_i}{dt} = \frac{\partial H}{\partial p_i}, \qquad \frac{dp_i}{dt} = -\frac{\partial H}{\partial q_i} \tag{3.15}$$

The time variation of the variables $q_i(t), p_i(t)$ is a trajectory in the phase space (q, p).

The standard method of integrating such equations is to make a Taylor expansion of each variable in a time interval δt as follows:

$$\begin{aligned} q_i(t + \delta t) &= q_i(t) + q_i'(t)\delta t + \tfrac{1}{2}q_i''(t)\delta t^2 + \cdots \\ p_i(t + \delta t) &= p_i(t) + p_i'(t)\delta t + \tfrac{1}{2}p_i''(t)\delta t^2 + \cdots \end{aligned} \tag{3.16}$$

The highest derivative retained in the expansion defines the order of the method. The first-order solution is trivial; using (3.15) we have

$$q_i(t+\delta t) = q_i(t) + \frac{\partial H}{\partial p_i}\delta t$$
$$p_i(t+\delta t) = p_i(t) - \frac{\partial H}{\partial q_i}\delta t \qquad 3.17$$

For higher orders there are various methods for estimating the higher derivatives in (3.16). For example, a second derivative can be obtained from the change in first derivatives at successive time intervals. It is not necessary to go deeply into this topic to see that the computational effort required to integrate Hamilton's equations increases with two factors: the first is the number of variables (p, q) and the second is the rate of change of these variables. Now although we shall see later (Chapter 5) that the choice of variable is not as critical in classical mechanics as it is in quantum mechanics, a well-chosen set of variables can save significantly on the computational demands of a problem. This is particularly true for vibrating molecules where some of the coordinates and momenta oscillate rapidly with time.

Change of variable is not a trivial matter in classical mechanics because Hamilton's equations (3.15) are only true if q_i and p_i are so-called canonical variables. The set of cartesian position variables (x_i, y_i, z_i) of particles i and their associated momenta

$$(p_{x_i}, p_{y_i}, p_{z_i}) = (m_i\dot{x}_i, m_i\dot{y}_i, m_i\dot{z}_i) \qquad 3.18$$

are canonical variables, but any change from these to another set of variables must be made according to the rules of classical mechanics. We will illustrate this with the simple example of a one-dimensional harmonic oscillator.

If q is the displacement of the oscillator from equilibrium the momentum is $p = m\dot{q}$ and the classical hamiltonian is

$$H = \left(\frac{p^2}{2m}\right) + \left(\frac{kq^2}{2}\right) \qquad 3.19$$

where m is the reduced mass of the oscillator and k the force constant.

Hamilton's equations are

$$\frac{dq}{dt} = \frac{p}{m}, \qquad \frac{dp}{dt} = -kq \qquad 3.20$$

and by substitution one can show that a solution of these equations is

$$q(t) = \left(\frac{2E}{k}\right)^{1/2} \sin(\omega t + \phi)$$
$$p(t) = (2Em)^{1/2} \cos(\omega t + \phi) \qquad 3.21$$

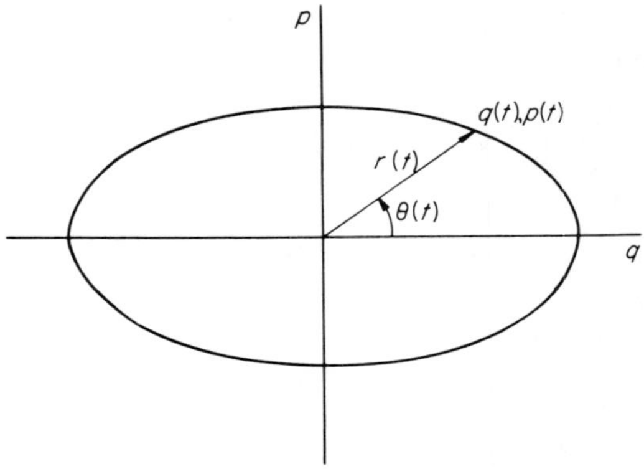

Figure 3.3 Classical path in phase space of a harmonic oscillator

where ω, the period of vibration, is given by

$$\omega = \left(\frac{k}{m}\right)^{1/2} \tag{3.22}$$

ϕ is an arbitrary phase, and E is the energy of the system. Figure 3.3 shows the trajectory in these variables, which is an ellipse whose major and minor axes are proportional to $E^{1/2}$. If one were solving equation (3.20) in this example by numerical integration of the Hamilton equations then in the first-order approximation we have, from (3.17),

$$\begin{aligned} q(t+\delta t) &= q(t) + \frac{p(t)\delta t}{m} \\ p(t+\delta t) &= p(t) - kq(t)\delta t \end{aligned} \tag{3.23}$$

At this level of approximation one would be replacing the ellipse of Figure 3.3 by a set of straight line segments; this is clearly a poor approximation unless δt is extremely small.

Not surprisingly there are alternative variables for the harmonic oscillator which are more amenable to numerical treatment. We could, for example, specify a point on the elliptic trajectory in polar coordinates

$$r(t) = (p^2 + q^2)^{1/2}, \qquad \theta(t) = \tan^{-1}\left(\frac{q}{p}\right) \tag{3.24}$$

and these would have a slower variation with time than either p or q separately. However, we shall now see that there are even better variables for this system.

From (3.21) we can anticipate that a good variable will be the angle α defined by

$$\alpha = \tan^{-1}\left(\frac{m^{1/2}k^{1/2}q}{p}\right) = \omega t + \phi \tag{3.25}$$

because α [unlike θ in (3.24)] is strictly a linear function of time. We now have to find the conjugate variable to this angle, and there is a standard procedure in classical mechanics for doing this.

The transformation from one set of canonical variables (P, Q) to another (P', Q') is called a canonical or contact transformation. Knowing (P, Q) and either P' and Q' we find the fourth variable (Q' or P' respectively) by constructing a generating function F whose partial differential gives the required variable.† The particular generating function that allows us to find the conjugate variable P', given P, Q and Q', is $F_1(Q, Q')$, and it satisfies the equations

$$P = \frac{\partial F_1}{\partial Q}, \qquad P' = -\frac{\partial F_1}{\partial Q'} \tag{3.26}$$

From (3.25) we see that

$$p = m^{1/2}k^{1/2}q \cot \alpha \tag{3.27}$$

and this is clearly given by the generating function

$$F_1(q, \alpha) = \tfrac{1}{2}m^{1/2}k^{1/2}q^2 \cot \alpha \tag{3.28}$$

because

$$p = \frac{\partial F_1(q, \alpha)}{\partial q} \tag{3.29}$$

The conjugate variable to α, which we call N, is then obtained from (3.26) as

$$N = -\frac{\partial F_1(q, \alpha)}{\partial \alpha} = \tfrac{1}{2}m^{1/2}k^{1/2}q^2 \operatorname{cosec}^2 \alpha \tag{3.30}$$

The angle α can be eliminated from expressions (3.30) and (3.25) to give

$$N\omega = \left(\frac{p^2}{2m} + \frac{kq^2}{2}\right) \tag{3.31}$$

and as the right-hand side is the hamiltonian (3.19) we see that N is a constant of the system given by

$$N = \frac{E}{\omega} \tag{3.32}$$

The hamiltonian in the new variables (N, α) can easily be obtained by finding p and q in terms of α and N and substituting these expressions into (3.19). The result is

$$H' = \omega N(\cos^2\alpha + \sin^2\alpha) = \omega N \tag{3.33}$$

so that H' is independent of the angle variable. Hamilton's equations in the new variables are trivially

$$\dot{\alpha} = \left[\frac{\partial H'}{\partial N}\right] = \omega, \qquad \dot{N} = -\left[\frac{\partial H'}{\partial \alpha}\right] = 0 \tag{3.34}$$

†See, for example, H. Goldstein, *Classical Mechanics*, Ch. 8, Addition-Wesley, New York, 1959.

Now although there is clearly no need to employ numerical integration of (3.34) to find the trajectory for the harmonic oscillator, it will be needed, in general, for collision problems involving anharmonic oscillators. If we defined the variables (N, α) for such systems using the same generating function F_1 (3.28) then N would be a slowly varying function of time and α will still be approximately linear in time. We would therefore find it easier to integrate Hamilton's equations in these variables than in the variables p and q. It is even possible that for an anharmonic oscillator one can find other variables similar to N and α such that one of these is strictly a constant of the motion.

The variables (N, α) defined above are called action-angle variables. The classical action (1.33) taken for one complete cycle of the oscillator trajectory is written

$$\oint p \, \mathrm{d}q \tag{3.35}$$

This is also the area of the ellipse which is $(2\pi E/\omega)$; from (3.32),

$$\oint p \, \mathrm{d}q = 2\pi N \tag{3.36}$$

Thus from (3.32) the variable N is proportional to the semi-classical quantum number for the vibrational state of the oscillator. Action-angle variables therefore have clear advantages over cartesian variables if we are applying semi-classical methods to the dynamics of oscillating systems.

Indeed, the semi-classical wave function [e.g. (2.68)] diverges when $p = 0$ (at a classical turning point) whereas in action-angle variables there is no such divergence.

The most commonly studied A + BC collinear collision system with realistic potentials is one in which there is an exponential repulsion between the colliding atoms

$$V_{\mathrm{AB}} = K \exp[-a(x_{\mathrm{B}} - x_{\mathrm{A}})] \tag{3.37}$$

A typical relationship between the energy transfer and the phase of the oscillator is shown in Figure 3.4 with parameters selected to model a collision between He and H_2. For this figure we take $t = 0$ to be the start of the classical trajectory and not, as in Figure 3.2, the phase at the time of collision as this, for a soft collision, is not a definable quantity.

The average energy transferred to the oscillator is

$$\Delta E = \frac{1}{2\pi} \int_0^{2\pi} \Delta E(\alpha) \, \mathrm{d}\alpha \tag{3.38}$$

This quantity combines both gains and losses from the oscillator. Of more interest is the change of vibrational state on collision and we can deduce this by using a semi-classical quantization of the oscillator vibrational levels.

Quantization is introduced into classical mechanics by assigning specific values to the action integral. For the harmonic oscillator we have already

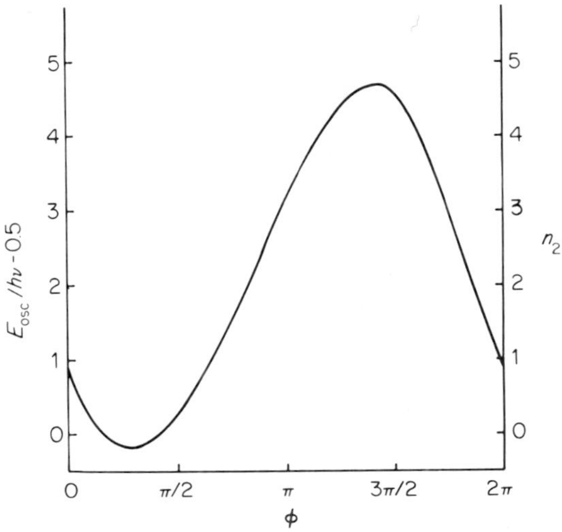

Figure 3.4 Relationship between the energy transfer and the oscillator phase for a soft collinear collision with parameters chosen to model $He + H_2$. The initial quantum number is $n_1 = 1$

derived an expression in (3.36) between this integral and the action variable N. There is therefore a direct relationship between N and the vibrational quantum number, n, which is

$$(n + \tfrac{1}{2})h = 2\pi N \qquad 3.39$$

As the energy of the harmonic oscillator is given by

$$E_n = (n + \tfrac{1}{2})h\nu_0 \qquad 3.40$$

where ν_0 is the frequency of the fundamental band, we can deduce the value of n after the collision by looking at the asymptotic value of the vibrational energy.

There are two ways of identifying the final vibrational quantum number from the results shown in Figure 3.4. The first, called the quasi-classical approach, is to assign any final state in which the value of n lies between $k - \frac{1}{2}$ and $k + \frac{1}{2}$, where k is an integer, to the quantum state k. Thus if trajectories are calculated from a random sample of the oscillator phase the probability of finding the system in the final state k is given by

$$\frac{N_k}{\sum_k N_k} \qquad 3.41$$

where N_k is the number of trajectories whose final quantum number is, by the above recipe, k.

The second approach is to use the semi-classical quantization of the oscillator and select only those trajectories that lead to a final integer value of n. For example, it is seen in Figure 3.4 that the final state $n = 2$ is obtained at only two values of the phase.

If all trajectories from an initial state n_1 with phases lying between α_1 and $\alpha_1 + \delta\alpha_1$ lead to a final quantum number between n_2 and $n_2 + \delta n_2$ then the contribution to the probability from this set of trajectories is

$$\frac{1}{2\pi}\left(\frac{\delta\alpha_1}{\delta n_2}\right) \tag{3.42}$$

In the limit of $\delta\alpha_1$ and δn_2 approaching zero we can replace the ratio in (3.42) by a derivative and hence obtain the probability for the transition $n_1 \rightarrow n_2$ as

$$P_{n_2 n_1} = \sum \left[2\pi \left(\frac{\mathrm{d}n_2}{\mathrm{d}\alpha_1} \right) \right]^{-1} \tag{3.43}$$

where the summation includes all trajectories that lead from n_1 to n_2. The derivative in (3.43) is a function of the initial-state quantum number n_1. For the example shown in Figure 3.4 this formula gives a transition probability $P_{21} = 0.13$

A classical trajectory that leads from n_1 to n_2 will, under time reversal, be coincident with that which goes from n_2 to n_1. It follows that expression (3.43) gives identical values for $P_{n_2 n_1}$ and $P_{n_1 n_2}$. The quasi-classical method, which is based on expression (3.41), does not have this symmetry, because trajectories that start in a discrete n_1 end up in the band $n_2 \pm \frac{1}{2}$. However, the quasi-classical method has the advantage in predicting small but non-zero transition probabilities for open channels when the state-to-state trajectory is classically forbidden. For example, it can be seen from Figure 3.4 that there is no trajectory leading from $n_1 = 1$ to $n_2 = 5$. However, there are trajectories that lead to values of n_2 greater than 4.5 and in the quasi-classical method these would contribute to P_{15} in expression (3.41). By this measure P_{15} is calculated to be 0.12 which, as we will see later (Table 3.1), is reasonably close to the quantum mechanical result of 0.077.

Expression (3.43) clearly fails if a point of zero slope in the function $n_2(\alpha_1, n_1)$ coincides with an integer value of n_2. The failure is analagous to the rainbow singularity in elastic cross-sections (2.23). The proper treatment of such transitions requires quantum theory or a more elaborate semi-classical theory. Expression (3.43) also fails to allow for interference between two or more trajectories that lead to the same final quantum number (see the discussion in Sections 1.4 and 2.4). We will return to both of the above points after we have examined the quantum theory of the collinear atom–diatomic molecule collision in Section 3.5.

3.4 The multichannel equations

For atom–atom scattering the form of the continuum wave functions can be easily obtained by numerical integration of the Schrodinger equation (2.36). The partial wave expansion provides a mathematically convenient method of writing the scattering wave function as a sum of products of functions of the distance between the atoms r and functions of the angular variables, and there are numerical methods for deriving the phase shift δ_l (2.44) directly. However, as we have emphasized earlier, mathematical convenience does not necessarily mean that the partial wave expansion is rapidly convergent.

For molecular scattering the wave function has greater dimensionality; it depends on the vector **r** of relative motion of the two centres of mass and on a set of internal variables **s** which are associated with the vibrational and rotational states of the molecules. As **r** approaches infinity the interaction potential $V(\mathbf{r}, \mathbf{s})$ approaches zero and the wave functions can be written as products of continuum functions for the relative motion and the wave functions for the discrete internal states. It is not possible to obtain the asymptotic form of the scattering wave function by direct numerical integration of the many-dimensional Schrodinger equation with this boundary condition. The reason for this is not only that many-dimensional numerical integration is computationally time consuming but that we have to solve a bound-state problem in the internal variables **s**, within the overall energy continuum.

The procedure we follow is that which is always used to find the bound states of many-dimensional systems with non-separable potentials, namely to make an expansion of the wave function in a set of known functions, which is called the basis set. For bound-state problems we normally take a basis which extends over all variables, whereas for scattering wave functions it covers only the internal variables **s**. Using such a basis the many-dimensional Schrodinger equation can be reduced to a coupled set of equations in the vector **r**, which are called the multichannel equations; these equations can be further reduced by a partial wave expansion so that we finally integrate a set of coupled equations in the scalar variable r.

In broad terms there are two types of basis set for the internal states: sets that depend only on the internal variables **s** and sets that depend on both r and **s**. The former leads to what is called the diabatic representation of the wave function and the latter to an adiabatic representation.† Adiabatic bases are more complicated functions and may be difficult to obtain, but to compensate for this one should need fewer of them to provide a specified computational accuracy.

Both the diabatic and adiabatic sets must represent the eigenfunctions for the internal motion of the molecules in the asymptotic region. Thus if the diabatic set are the functions $\chi_j(\mathbf{s})$ the adiabatic set $\Lambda_j(r, \mathbf{s})$ will satisfy the condition

$$\lim_{r \to \infty} \Lambda_j(r, \mathbf{s}) = \chi_j(\mathbf{s}) \qquad 3.44$$

To derive the multichannel equations we write the total hamiltonian for the nuclear motion as a sum of three parts: the kinetic energy operator for the relative motion of the two centres of mass, the interaction potential and the hamiltonian

†The prefix dia (Greek through) and adia (not through) are adopted from the type of functions used to calculate potential energy curves and refer to the behaviour (crossing or non-crossing) of such curves.

for the internal motion H^0:

$$H = -\frac{\hbar^2}{2\mu}\nabla^2(\mathbf{r}) + V(r, \mathbf{s}) + H^0(\mathbf{s}) \tag{3.45}$$

$V(r, \mathbf{s})$ will approach zero as r approaches infinity. Note that the interaction potential depends only on the scalar distance between the interacting species and not on the angular components of $\mathbf{r}$ because these angular components describe the scattering angle for the centres of mass. The interaction potential is of course dependent on the orientations of the molecules relative to $\mathbf{r}$ but these angles can be considered as part of the set of internal coordinates $\mathbf{s}$.

In the diabatic basis we expand the total wave function as

$$\Psi(\mathbf{r}, \mathbf{s}) = \sum_j \psi_j(\mathbf{r})\chi_j(\mathbf{s}) \tag{3.46}$$

where $\chi_j(\mathbf{s})$, the eigenstates of H^0, are an orthonormal set satisfying

$$H^0(\mathbf{s})\chi_j(\mathbf{s}) = E_j^0\chi_j(\mathbf{s}) \tag{3.47}$$

The Schrodinger equation then takes the form

$$\sum_j\left[\frac{-\hbar^2}{2\mu}\nabla^2(\mathbf{r}) + V(r, \mathbf{s}) + H^0(\mathbf{s}) - E\right]\psi_j(\mathbf{r})\chi_j(\mathbf{s}) = 0 \tag{3.48}$$

which from (3.24) can be written as

$$\sum_j\left[\frac{-\hbar^2}{2\mu}\nabla^2(\mathbf{r}) + V(r, \mathbf{s}) + E_j^0 - E\right]\psi_j(\mathbf{r})\chi_j(\mathbf{s}) = 0 \tag{3.49}$$

This can be converted to a set of coupled equations in the variable $\mathbf{r}$ by multiplying in turn by each of the basis functions and integrating over $\mathbf{s}$. A typical equation is

$$\sum_j\left[\left\langle \chi_i(s)\middle| -\frac{\hbar^2}{2\mu}\nabla^2(\mathbf{r}) + E_j^0 - E \middle| \chi_j(s)\right\rangle + \langle\chi_i(s)|V(r, \mathbf{s})|\chi_j(s)\rangle\right]\psi_j(\mathbf{r}) = 0 \tag{3.50}$$

which from the orthogonality of the basis becomes

$$\left[-\frac{\hbar^2}{2\mu}\nabla^2(\mathbf{r}) + E_i^0 - E + V_{ii}(r)\right]\psi_i(\mathbf{r}) + \sum_{j\neq i} V_{ij}(r)\psi_j(\mathbf{r}) = 0 \tag{3.51}$$

where

$$V_{ij}(r) = \langle\chi_i(\mathbf{s})|V(r, \mathbf{s})|\chi_j(\mathbf{s})\rangle \tag{3.52}$$

More concisely, the set of equations can be written in matrix form as

$$\left[(\mathbf{E}^0 - \frac{\hbar^2}{2\mu}\nabla^2(\mathbf{r})\mathbf{I} - E\mathbf{I} + \mathbf{V}(r)\right]\boldsymbol{\psi}(\mathbf{r}) = \mathbf{0} \tag{3.53}$$

where $\boldsymbol{\psi}(r)$ is a column vector, $\mathbf{V}$ is the matrix whose elements are defined by (3.52), $\mathbf{I}$ is a unit matrix, and $\mathbf{E}^0$ is a diagonal matrix whose elements are the eigenvalues E_j^0.

It can be seen from (3.51) that in the diabatic basis the equations can be considered as a set of elastic equations, one for each basis function, which are coupled together by off-diagonal elements of the interaction potential $V_{ij}(r)$. Each equation is called a channel equation, V_{ii} and E_i^0 being called the channel potentials and channel energies respectively. Equations (3.53) are called the multichannel equations.

We now contrast this with the result obtained from an adiabatic basis. The simplest form of the equations is obtained if the basis functions are taken to be eigenfunctions of $H^0(\mathbf{s}) + V(r, \mathbf{s})$ as follows:

$$[H^0(\mathbf{s}) + V(r, \mathbf{s})]\Lambda_j(r, \mathbf{s}) = E_j(r)\Lambda_j(r, \mathbf{s}) \qquad 3.54$$

Note that the variable r appears as a parameter in the eigenfunction and eigenvalue so we must solve the equation for each value of r. This is very much in the spirit of the Born–Oppenheimer model in which electronic eigenfunctions and eigenvalues are obtained with the internuclear variables as parameters, and is the reason why Λ_j, like Born–Oppenheimer functions, is referred to as the adiabatic basis.

We now substitute

$$\Psi(\mathbf{r}, \mathbf{s}) = \sum_j \psi_j(\mathbf{r})\Lambda_j(r, \mathbf{s}) \qquad 3.55$$

into the Schrodinger equation with the hamiltonian (3.45) and form the matrix elements with the basis functions [multiplying by $\Lambda_i(r, \mathbf{s})$ and integrating over $\mathbf{s}$]. There are two differences from the diabatic equations. Firstly, there are no off-diagonal elements of $V(r, \mathbf{s})$ because the adiabatic basis are eigenfunctions of H^0 + $V(r, \mathbf{s})$ [the diagonal elements of $V(r, \mathbf{s})$ are subsumed into $E_j(\mathbf{r})$] and, secondly, there *are* off-diagonal elements of the kinetic energy operator $\nabla^2(\mathbf{r})$ because the adiabatic basis is parametrically a function of r. The equations in expanded form [cf. (3.51)] are

$$\left[\frac{-\hbar^2}{2\mu}\nabla^2(\mathbf{r}) + E_i(r) - E\right]\psi_i(\mathbf{r}) - \frac{\hbar^2}{2\mu}\sum_j\left[\nabla^2_{ij} + 2\nabla_{ij}\nabla(\mathbf{r})\right]\psi_j(\mathbf{r}) = 0 \qquad 3.56$$

where

$$\nabla^2_{ij} = \langle\Lambda_i(r, \mathbf{s})|\nabla^2(\mathbf{r})|\Lambda_j(r, \mathbf{s})\rangle \qquad 3.57$$

is the matrix of $\nabla^2(\mathbf{r})$, the kinetic energy operator for the relative motion of the two systems. The coupling in the adiabatic basis can therefore be said to arise through kinetic energy rather than potential energy terms.

Both the diabatic and adiabatic expansions provide, in principle, an exact description of the wave function; in quantum mechanical terms we say the bases are complete sets. However, in practice we must take a finite rather than an infinite number of terms in the expansions. The finite set of multichannel equations is usually called the close-coupling equations.

The important questions are, firstly, how many terms are needed to obtain a required level of accuracy and, secondly, how great is the computational task to reach this accuracy. Kinetic energy coupling terms (i.e. off-diagonal elements of ∇ and ∇^2) are expected to be smaller than potential energy coupling terms, because the basis functions are generally slowly varying functions of r. One therefore usually requires fewer terms in the adiabatic basis than in the diabatic basis. However, to obtain the adiabatic basis we must solve (3.54) over the range of r of interest, and that can be a formidable task. For this reason most calculations are made in the diabatic basis.

An alternative strategy, which may appear to combine the worst of both bases, is to choose, by physical intuition or a simple model, a basis which is a function of r but not necessarily an eigenfunction of $H^0(\mathbf{s}) + V(r, \mathbf{s})$. In this case there will be coupling terms from both the potential and kinetic energy operators, but with a judicious choice of basis both may be small. This can be the starting point of approximations in which small coupling terms are taken as zero.

To progress further with the coupled channel equations for full three-dimensional scattering we have to make a partial wave expansion of $\psi(\mathbf{r})$. As we shall see later this expansion, which can be made in more than one way, leads to rather complicated equations which reflect the fact that the orbital angular momentum of one molecule around the other is coupled with the rotational angular momenta of the molecules ('internal' motions), only the total of these being conserved. We can, however, illustrate the multichannel approach by a one-dimensional system in which no partial wave decomposition is required, i.e. $\mathbf{r}$ is a scalar rather than a vector, and this we now do for the collinear collision of an atom with a diatomic harmonic oscillator, the system studied in Section 3.2 by classical mechanics.

3.5 Quantum mechanical treatment of collinear atom–diatomic molecule collisions

In Section 3.2 we looked at the classical mechanics of the collinear collision between an atom and a diatomic molecule. We solved explicitly a simple hard sphere collision problem and examined the results for soft collisions. We now look at the quantum mechanical treatment of this problem and will follow this with a more thorough look at the semi-classical treatment.

The hamiltonian for an atom A colliding along the x axis with a harmonic diatomic oscillator BC of force constant k is as follows:

$$\mathbf{H} = -\frac{\hbar^2}{2m_A}\frac{\partial^2}{\partial x_A^2} - \frac{\hbar^2}{2m_B}\frac{\partial^2}{\partial x_B^2} - \frac{\hbar^2}{2m_C}\frac{\partial^2}{\partial x_C^2} + \frac{k}{2}(x_{BC} - x_{BC}^0)^2 + V_{A-BC} \qquad 3.58$$

where $x_{BC} = x_B - x_C$, and x_{BC}^0 is the equilibrium bond length of BC. By transferring to a centre-of-mass coordinate system

$$X_{CM} = \frac{x_A m_A + x_B m_B + x_C m_C}{M}$$

$$X = x_A - \frac{x_B m_B + x_C m_C}{m_B + m_C}$$

$$Y = x_B - x_C$$

$$M = m_A + m_B + m_C \qquad 3.59$$

the hamiltonian becomes

$$\mathbf{H} = \frac{-\hbar^2}{2M}\frac{\partial^2}{\partial X_{\mathrm{CM}}^2} - \frac{\hbar^2}{2\mu}\frac{\partial^2}{\partial X^2} - \frac{\hbar^2}{2\mu_{\mathrm{BC}}}\frac{\partial^2}{\partial Y^2} + \frac{k}{2}(Y - Y^0)^2 + V_{\mathrm{A-BC}} \qquad 3.60$$

where

$$\mu = \frac{m_{\mathrm{A}}(m_{\mathrm{B}} + m_{\mathrm{C}})}{M}, \qquad \mu_{\mathrm{BC}} = \frac{m_{\mathrm{B}} m_{\mathrm{C}}}{m_{\mathrm{B}} + m_{\mathrm{C}}} \qquad 3.61$$

As the centre-of-mass motion is uncoupled from the motion in X and Y, and is therefore conserved, the first term in (3.60) can be ignored. The X and Y motions are coupled through the interaction potential $V_{\mathrm{A-BC}}$. Indeed if this was a function only of X, the distance between A and the centre of mass of BC, then even the X and Y motions would be uncoupled. More realistically, however, the interaction potential is a function of the A – B distance which is

$$x_{\mathrm{A}} - x_{\mathrm{B}} = X - \frac{m_{\mathrm{C}} Y}{m_{\mathrm{B}} + m_{\mathrm{C}}} \qquad 3.62$$

The coordinates X and Y in (3.60) fill the roles of $\mathbf{r}$ and $\mathbf{s}$ respectively in (3.45), and $V_{\mathrm{A-BC}}$ represents the coupling potential $V(r, \mathbf{s})$. The diabatic basis functions (3.47) are the harmonic oscillator eigenfunctions

$$\chi_j(Y) = \left[\frac{k^{1/4}\mu_{\mathrm{BC}}^{1/4}}{\hbar^{1/2}\pi^{1/2}2^j j!}\right]^{1/2} H_j(Q)\exp\left(\frac{-Q^2}{2}\right) \qquad 3.63$$

where

$$Q = \left(\frac{k\mu_{\mathrm{BC}}}{\hbar^2}\right)^{1/4}(Y - Y^0) \qquad 3.64$$

and $H_i(Q)$ are Hermite polynomials. The basis set eigenvalues are

$$E_j^0 = \left(\frac{\hbar^2 k}{\mu_{\mathrm{BC}}}\right)^{1/2}\left(j + \frac{1}{2}\right) \qquad 3.65$$

The multichannel equations for the system in the diabatic basis (3.51) are

$$\left[-\frac{\hbar^2}{2\mu}\frac{\mathrm{d}^2}{\mathrm{d}X^2} + E_i^0 - E + V_{ii}(X)\right]\psi_i(X) + \sum_{j \neq i} V_{ij}(X)\psi_j(X) = 0 \qquad 3.66$$

where

$$V_{ij}(X) = (2^{i+j}\pi i! j!)^{-1/2}\int_{-\infty}^{\infty} H_i(Q)H_j(Q)\mathrm{e}^{-Q^2}V\left(X - \frac{m_{\mathrm{C}} Y^0}{m_{\mathrm{B}} + m_{\mathrm{C}}} - \varepsilon Q\right)\mathrm{d}Q \qquad 3.67$$

and

$$\varepsilon = \frac{m_{\mathrm{C}}}{m_{\mathrm{B}} + m_{\mathrm{C}}}\left(\frac{\hbar^2}{k\mu_{\mathrm{BC}}}\right)^{1/2} \qquad 3.68$$

The set of equations (3.66) have only one variable (X) and can be solved using the standard numerical techniques which are available for ordinary differential equations. These involve numerical integration outwards from $X = 0$ where the interaction potential $V_{\mathrm{A-BC}}$ is infinitely repulsive and the channel wave functions must all be zero. There are several methods for carrying out this integration and those currently most widely used are discussed briefly in Appendix 4.

To start the integration it is also necessary to specify initial values for the derivatives $\psi_i'(0)$. These can initially be taken as arbitrary, although we shall see shortly the solutions are eventually transformed to satisfy the boundary conditions for the problem.

The multichannel equations must be integrated outwards to a sufficiently large value of X that all the elements $V_{ij}(X)$ are negligible. At this point the equations have the form

$$\left(-\frac{\hbar^2}{2\mu}\frac{d^2}{dX^2} + E_i^0 - E\right)\psi_i(X) = 0 \qquad 3.69$$

and the general solutions of these are therefore the asymptotic solutions to the multichannel equations

$$\psi_i(X) \sim A_i e^{ik_iX} + B_i e^{-ik_iX} \qquad 3.70$$

where

$$k_i = \left[\frac{2\mu(E - E_i^0)}{\hbar^2}\right]^{1/2} \qquad 3.71$$

are the channel wave numbers. The coefficients A_i and B_i are called Jost functions; their values depend on the choice of the initial derivatives $\psi_i'(0)$.

If N diabatic basis functions (3.63) are used to expand the scattering wave function we obtain N multichannel second-order ordinary differential equations (3.66) and for each value of the energy (E) there are $2N$ linearly independent solutions, N of which can be chosen to satisfy the boundary condition $\psi_i(0) = 0$. The simplest way of constructing such a set is to use the following recipe: for the nth solution take $\psi_n'(0)$ to be arbitrary and all other $\psi_i'(0)$ $(i \neq n)$ to be zero. The nth solution will produce Jost functions A_{in} and B_{in}, so that the full set of Jost functions are matrices **A** and **B**.

Returning now to (3.71) we see that the channel wave numbers are real for $E > E_i^0$ and imaginary for $E < E_i^0$; by convention the latter are always taken as positive imaginary. When $E > E_i^0$ it is possible for the system to emerge from the scattering as A + BC because the total energy is greater than the internal energy of BC; these are called open channels. When $E < E_i^0$ it is not physically possible to produce A + BC because the total energy is less than that contained in the internal state of the oscillator; these are called closed channels.

For closed channels the asymptotic form of the wave function (3.70) will be an exponentially increasing function of X (ik_i is real) unless $B_i = 0$. This is therefore a boundary condition which must be imposed on the scattering wave functions to ensure that closed channels are not produced at the asymptotic limit.

If we want a wave function that represents the system in an initial internal state i, emerging in final states j which are open channels, then we write the jth component as

$$\psi_j(X) \sim \delta_{ij} e^{-ik_iX} - S_{ji}\left(\frac{k_i}{k_j}\right)^{1/2} e^{ik_jX} \qquad 3.72$$

where the coefficients S_{ji} are elements of a matrix which is called the scattering matrix or **S** matrix. The Kronecker delta, δ_{ij}, becomes in matrix form the unit matrix. The factor $(k_i/k_j)^{1/2}$ is introduced so that the square modulus of **S** gives fluxes and not just probability densities [see (2.22)]. The details of the transformation by which the Jost matrices of (3.70) are related to the **S** matrix are given in Appendix 3. We also leave until Section 4.4 a discussion of the properties of **S** and its relationship to other important matrices used in scattering theory.

Secrest and Johnson† have solved the coupled channel equations for the collinear collision between an atom and a harmonic oscillator using the exponential interaction potential of the type given by (3.37). They showed that if the collision energy was expressed in units of the oscillator energy $h\nu_0$, then the results depend on only two parameters, one of which was an exponent α in the interaction potential (their distances are scaled by parameters that depend on both the masses and the force constant of the molecule) and the other a mass ratio

$$m = \frac{m_A m_C}{m_B(m_A + m_B + m_C)} \qquad 3.73$$

The channel potentials (3.67) $V_{ii}(X)$ are exponentially decreasing functions of X, reflecting the exponential repulsion between the colliding atoms, and their magnitudes are all rather similar. The off-diagonal elements $V_{ij}(X)$ are also exponentially decreasing but their magnitude gets smaller as the difference $|n_i - n_j|$ increases. This is because the product of the diabatic basis functions $\chi_i(Y)$ and $\chi_j(Y)$ has an increasing number of sign changes in the Y dimension as $|n_i - n_j|$ increases.

Results are shown in Table 3.1 for $\alpha = 0.3$ and $m = \frac{2}{3}$ to simulate a $He-H_2$ collision; this is a rather 'soft' collision. These parameters were also used for the classical calculations described earlier (Figure 3.4). The table includes results obtained by Miller from classical trajectories using semi-classical quantization of the oscillator and by a more elaborate semi-classical theory which we will be discussing in the next section.

The transition probabilities are symmetric in the initial and final state quantum numbers for a fixed total energy. This, called the principle of detailed balance, is connected with the symmetry of the equations of motion under time reversal and will be discussed in Section 4.4.

The only specific point we make at this time about the results of Table 3.1 is that the quantum mechanical transition probabilities are small for those open channels for which there is no classical trajectory leading from n_1 to n_2 [hence a zero probability from expression (3.43)].

Figure 3.5 shows the calculated transition probabilities for this system from the state $n = 2$ to other states as a function of the collision energy. The system becomes increasingly inelastic as the total energy is increased. This is mainly because an increasing number of channels become open, and although an

†D. Secrest and B. R. Johnson, *J. Chem. Phys.*, **45**, 4556 (1966).

Table 3.1 Transition probabilities for He–H_2 collisions at an energy of $10h\nu_0$ ($\sim$4.5 eV) with collision parameters $\alpha = 0.3$, $m = \frac{2}{3}$.

Transition		Quantum*	Classical*	Semi-classical†	
n_1	n_2			Simple	Uniform
0	0	0.060	0	0	0.058
0	1	0.218	0.356	0.472	0.211
0	2	0.366	0.212	0.416	0.381
0	3	0.267	0.232	0.359	0.266
0	4	0.089	0	0	0.075
1	1	0.286	0.158	0.290	0.287
1	2	0.009	0.130	0.009	0.011
1	3	0.170	0.128	0.168	0.174
1	4	0.240	0.159	0.285	0.240
1	5	0.077	0	0	0.062
2	2	0.366	0.212	0.416	0.381
2	3	0.018	0.105	0.020	0.017
2	4	0.169	0.114	0.165	0.170
2	5	0.194	0.169	0.262	0.194
2	6	0.037	0	0	0.045

*References to this work have been given earlier.
†The semi-classical results will be discussed in Section 3.6.

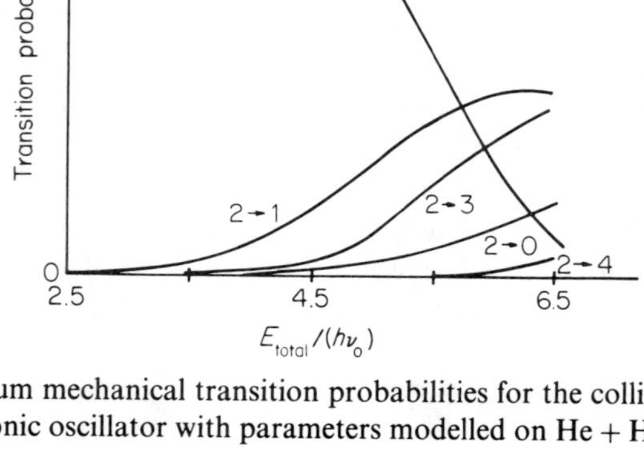

Figure 3.5 Quantum mechanical transition probabilities for the collinear collision of an atom and a harmonic oscillator with parameters modelled on He + H_2 ($\alpha = 0.314$, $m = \frac{2}{3}$)

individual inelastic transition probability does not necessarily increase with increasing energy, their total increases up to the maximum of one.

The collinear A + BC collision becomes increasingly inelastic as the value of α increases, i.e. as the collision becomes harder. For hard sphere collisions ($\alpha = \infty$), an individual inelastic transition probability increases very rapidly with energy from its threshold and then decreases rapidly from the threshold of the next channel. In other words, a new open channel 'steals' some transition probability from other open channels.

3.6 The semi-classical approach to inelastic collisions

In Section 3.3 we showed how semi-classical quantization of the vibrational states of a diatomic molecule could be introduced through the use of action-angle variables and how state-to-state transition probabilities could be calculated for the collinear A + BC collision. The formulae used for the transition probabilities (3.41) and (3.43) do not, however, allow for interference between trajectories. In this section we show how interferences can be introduced by using the semi-classical method to calculate a transition amplitude whose squared modulus will give the transition probability.

We have seen earlier that a phase can be associated with each classical trajectory by evaluating the action integral (1.30). If there is more than one coordinate and conjugate momentum, the classical action is

$$A = \sum_k \int_i^f p_k \, \mathrm{d}q_k \tag{3.74}$$

where i and f indicate that the limits of the integral are the initial and final points of the trajectory. As trajectories are calculated by integrating Hamilton's equations over time (3.15) it is convenient to calculate the action in the same way by changing the integration variable to give the expression

$$A = \int_i^f \mathrm{d}t \sum_k p_k \dot{q}_k \tag{3.75}$$

For systems with only one coordinate variable the JWKB expansion of the wave function provides a rigorous link between the classical action and the phase of the semi-classical wave function, but for systems of higher dimension we can only postulate that a similar link exists; there is no exact multidimensional JWKB solution. There have been two distinct approaches to the problem. One proposed by Marcus extends the JKWB approach to many dimensions and the other, due mainly to Miller, follows a path integral approach to quantum mechanics which was developed by Feyman and Hibbs.† The two methods lead to the same outcome, and both make substantial use of classical mechanics, particularly of

†R. P. Feynman and A. R. Hibbs, *Quantum Mechanics and Path Integrals*, McGraw-Hill, New York, 1965.

the canonical variable transformations which were touched on briefly in Section 3.3. We give only the outline of the theory here and direct more demanding readers to fuller discussions of the problem.†

The first point to emphasize is that by using action-angle variables for the oscillator we avoid singularities in the semi-classical wave function that occur at the classical turning points ($p = 0$). Although there may also be classical turning points for the scattering coordinate they give no difficulty because they occur far away from the asymptotic regions where the transition probabilities are determined.

The action-angle equivalent of (3.74) is

$$\int_i^f N\,\mathrm{d}\alpha \tag{3.76}$$

For an isolated harmonic oscillator N is a constant of the motion from (1.30) and (3.39) this gives a phase of

$$(n + \tfrac{1}{2})(\alpha_f - \alpha_i) \tag{3.77}$$

However, for inelastic collisions of an atom and a harmonic oscillator, N is a variable of the motion so the integral (3.76) must be evaluated along the trajectory.

For the collinear A + BC problem, the variables are X, the scattering coordinate (3.59), and its conjugate momentum $P = \mu\dot{X}$, and α (3.25) and its conjugate action variable $N = (n + \frac{1}{2})h$ [(3.32) and (3.39)]. Trajectories describe the time variations of these variables [$X(t)$, $P(t)$, $\alpha(t)$, $N(t)$], and we seek trajectories that lead from specified initial values of N (given by n_1) to specified final values of N (given by n_2).

In the asymptotic regions P and N became constant whereas X and α still change with time. It is therefore clear that the value of the classical action, as given by

$$A = \int_i^f \mathrm{d}t(P\dot{X} + N\dot{\alpha}) \tag{3.78}$$

will depend on the precise values of X at which the trajectory is initiated and stopped; the final value of α also depends on the final value of X. However, integrating (3.76) by parts gives

$$A = [PX + N\alpha]_i^f - \int_i^f \mathrm{d}t(\dot{P}X + \dot{N}\alpha) \tag{3.79}$$

and the integral in this expression is independent of the initial and final value of X, because $\dot{P} = \dot{N} = 0$ in the asymptotic region where the interaction potential is zero. In fact Miller has shown that as the trajectories are specified by the initial and final values of the action variable, one requires the classical equivalent of the

†For example, W. H. Miller, *Adv. Chem. Phys.*, **25**, 69 (1974).

momentum representation of quantum mechanics rather than the more familiar coordinate representation. The appropriate phase for this kind of trajectory is the integral that appears in (3.79) whose value, as we have seen, is independent of the precise end points of the trajectory.

If we now include the phase in the classical expression for the inelastic transition probabilities (3.43), we obtain

$$P_{n_2 n_1} = \left| \sum_j \left[2\pi \left(\frac{dn_2}{d\alpha_j} \right) \right]^{-1} \exp(i\phi_j) \right|^2 \qquad 3.80$$

where

$$\phi_j = -\hbar^{-1} \int_i^f dt(\dot{P}X + \dot{N}\alpha) \qquad 3.81$$

the summation in (3.80) being over each trajectory leading from n_1 to n_2. For the example shown in Figure 3.4, there are two trajectories that lead to $n_2 = 2$ and hence the transition probability is

$$P_{n_1 n_2} = P_a + P_b + 2(P_a P_b)^{1/2} \sin(\phi_a - \phi_b) \qquad 3.82$$

where P_a and P_b are the classical probabilities, i.e. those calculated without the interference term. The interference term is a sine function rather than a cosine because the slopes of $(dn_2/d\alpha)$ have different signs for the two trajectories.

Table 3.1 shows some semi-classical transition probabilities calculated by Miller for the collinear A + BC collisions. Comparison with the classical results shows that the interference terms are quite large although the semi-classical values are not always in better agreement with the exact quantum mechanical values. Moreover, when there is no classical trajectory leading from n_1 to n_2, even when the channel is open, the semi-classical transition probability is still zero in this model. The semi-classical results become poorer as the classically forbidden region is approached and expression (3.80) breaks down completely if $(dn_2/d\alpha_j)$ is zero; the equivalent of the classical rainbow singularity. The semi-classical method can be improved, as it can for classical rainbows, by taking a higher level of approximation called the uniform approximation, in which Airy functions rather than the sine function are used for the interference term. With this development the interference term can also be analytically continued into the classically forbidden region, using the property that the Airy function decays exponentially for large positive values of the argument, as illustrated in Figure 2.17. It can be seen from Table 3.1 that the results with this model are in excellent agreement with the quantum results.

Another approach to classically forbidden transitions is to allow the trajectories to 'tunnel' through a classically forbidden region by integrating in the complex time domain; coming back into real time to find a real specified quantum number n_2. It is interesting to note that Hamilton's equations of motion give quite a good representation of tunnelling if the time and space variables, and hence momenta, are allowed to be complex. However, this is at the expense of doubling

†W. H. Miller, *J. Chem. Phys.*, **53**, 3578 (1970).

the number of independent variables because trajectories are defined by the time variation of both the real and imaginary parts of the coordinates and momenta.

Although the semi-classical method, particularly in the uniform model, gives excellent results for inelastic transition probabilities for the collinear A + BC collision, as it does also for the rotationally inelastic collisions of an atom and a rigid diatomic molecule, it becomes difficult to apply if there is a larger number of internal degrees of freedom. Every internal variable for periodic motion has an angle variable α and the method requires a search through the α space to find trajectories that lead from a specified initial set of action variables N_i to a specified final set N_f; this is not easy. Of course, the quantum mechanical approach is also not easy because one requires basis functions for each internal variable and the number of coupled-channel equations may become very large. For this reason, approximations to the exact quantum mechanical results are very important.

3.7 Approximate solutions of the coupled-channel equations

The main difficulty encountered in solving the coupled-channel equations is that a very large number of channels (or basis functions) are needed to obtain convergence in the transition probabilities, and hence the numerical integration of the equations, (3.51) or (3.56), is a task that strains the resources of even the largest computer. Approximate solutions also provide insight into the basic physics of the problem. There are a number of approximations that can be applied to the problem, all of which depend on some coupling terms in either the diabatic or adiabatic equations being small or negligible. For the collinear A + BC collision exact solutions of the quantum mechanical equations are now rather easy to obtain, but we must nonetheless discuss approximate methods because of their relevance to more difficult problems.

The first approximation we discuss is an iterative perturbation method based on an approach to continuum wave functions first proposed by Born in 1926. For inelastic scattering this is called the distorted wave Born approximation. If all off-diagonal elements of (3.52) are put to zero, equations (3.51) are uncoupled with each equation being of the form

$$\left[-\frac{\hbar^2}{2\mu}\nabla^2(\mathbf{r}) + \mathbf{E}_i^0 - E + V_{ii}(r)\right]\psi_i(\mathbf{r}) = 0 \qquad 3.83$$

The solution to this equation describes elastic scattering in a central potential $V_{ii}(r)$ with the colliding fragments being in an internal state i; these solutions we called distorted waves.

If $\psi_i^{(0)}$ is a solution of (3.74) then we can find an improved solution of (3.51) by writing

$$\psi_i = \psi_i^{(0)} + \psi_i^{(1)} \qquad 3.84$$

and inserting this into (3.51) and neglecting the terms $V_{ij}(r)\psi_i^{(1)}(r)$, which are assumed to be small, we obtain the following equation for $\psi_i^{(1)}$:

$$\left[-\frac{\hbar^2}{2\mu}\nabla^2(\mathbf{r})+\mathbf{E}_i^0-\mathbf{E}+V_{ii}(r)\right]\psi_i^{(1)}(\mathbf{r})=-\sum_{j\neq i}V_{ij}(r)\psi_j^{(0)}(r) \qquad 3.85$$

These equations are still uncoupled, i.e. they can be solved independently, because the terms on the right-hand side of (3.85) are $\psi_j^{(0)}(r)$ which are the distorted waves. The sequence leading from (3.83) to (3.85) is the first step in an iterative approach to the scattering wave functions which could be taken further by adding a further term $\psi_i^{(2)}$ to the wave function and repeating the cycle. A similar iterative procedure can be applied to the adiabatic equations (3.56).

For the collinear A + BC collision, equations (3.85) take the form

$$\left[-\frac{\hbar^2}{2\mu}\frac{\mathrm{d}^2}{\mathrm{d}X^2}+E_i^0-E+V_{ii}(X)\right]\psi_i^{(1)}(X)=-\sum_{j\neq i}V_{ij}(X)\psi_j^{(0)}(X) \qquad 3.86$$

However, as discussed earlier, from a basis of N internal states such equations have N linearly independent solutions with the boundary condition $\psi_i^{(1)}(0)=0$ and there are an unlimited number of ways of producing such solutions.

One way, the most convenient, is to retain only one term on the right-hand side and to take for the nth solution the set of equations

$$\left[-\frac{\hbar^2}{2\mu}\frac{\mathrm{d}^2}{\mathrm{d}X^2}+E_i^0-E+V_{ii}(X)\right]\psi_{in}^{(1)}(X)=\begin{cases}-V_{in}(X)\psi_n^{(0)}(X) & (i\neq n)\\ 0 & (i=n)\end{cases} \qquad 3.87$$

where we have added a suffix n to indicate this solution. Any solution of (3.86) is a linear combination of the set of solutions for each n.

Equations (3.87) can be rewritten

$$\frac{\mathrm{d}^2\psi_{in}^{(1)}(X)}{\mathrm{d}X^2}=[U_{ii}(X)-k_i^2]\psi_{in}^{(1)}(X)+U_{in}(X)\psi_n^{(0)}(X) \qquad 3.88$$

where

$$U_{in}(X)=\frac{2\mu}{\hbar^2}V_{in}(X) \qquad 3.89$$

This is a set of homogeneous linear equations which has a general solution

$$\psi^{(1)}(X)=\int_{X_0}^{X}K_i(X,X')U_{in}(X')\psi_n^{(0)}(X')\,\mathrm{d}X' \qquad 3.90$$

where $K_i(X,X')$ is called the kernel. X_0 is a point in the classically forbidden region where it can be assumed that $\psi_{in}^{(1)}(X)$ is zero. The kernel is given by

$$K_i(X,X')=\frac{1}{2ik_i}[f_i^+(X)f_i^-(X')-f_i^+(X)f_i^-(X')] \qquad 3.91$$

where f_i^+ and f_i^- are two irregular solutions of the homogeneous equation

$$\frac{\mathrm{d}^2f_i^\pm(X)}{\mathrm{d}X^2}=[U_{ii}(X)-k_i^2]f_i^\pm(X) \qquad 3.92$$

which are defined with boundary conditions

$$f_i^{\pm}(X) \sim \exp(\pm \mathrm{i}k_i X) \tag{3.93}$$

However, $f_i^{\pm}(X)$ are not zero at X_0, nor does the boundary condition at $X = \infty$ necessarily produce the required form of the wave function, which is (3.72). The boundary conditions are handled as for the full solution of the multichannel equations by the use of Jost functions (3.70). However, closed channels are not to be included because, for these, f_i^- would be exponentially divergent at large X.

Substituting (3.91) into (3.90) gives

$$\psi^{(1)}(X) = \frac{1}{2\mathrm{i}k_i}\left[f_i^+(X)\int_{X_0}^{X} f_i^-(X')U_{in}(X')\psi_n^{(0)}(X')\,\mathrm{d}X' - f_i^-(X)\int_{X_0}^{X} f_i^+(X')U_{in}(X')\psi_n^{(0)}(X')\,\mathrm{d}X' \right] \tag{3.94}$$

Taking the asymptotic limit and comparing with (3.70) gives the Jost matrices in first order as

$$\begin{aligned} A_{in}^{(1)} &= \frac{1}{2\mathrm{i}k_i}\int_{X_0}^{\infty} f_i^-(X)U_{in}(X)\psi_n^{(0)}(X)\,\mathrm{d}X \\ B_{in}^{(1)} &= \frac{1}{2\mathrm{i}k_i}\int_{X_0}^{\infty} f_i^-(X)U_{in}(X)\psi_n^{(0)}(X)\,\mathrm{d}X \end{aligned} \tag{3.95}$$

The zeroth-order terms being defined by the distorted wave functions [from (3.70)]

$$\psi_i^{(0)}(X) \sim A_{ii}^{(0)}\mathrm{e}^{\mathrm{i}k_i X} + B_{ii}^{(0)}\mathrm{e}^{-\mathrm{i}k_i X} \tag{3.96}$$

The **S** matrix will, from (3.70) and (3.72), be

$$S_{ji} = -\left(\frac{k_j}{k_i}\right)^{1/2}(AB^{-1})_{ji} \tag{3.97}$$

$A^{(0)}$ and $B^{(0)}$ are diagonal, because the scattering is elastic, so $S^{(0)}$ will be diagonal and equal to $A^{(0)}(B^{(0)})^{-1}$. The next term arises from combinations of $A^{(0)}$ with $B^{(1)}$ and $A^{(1)}$ with $B^{(0)}$. Expanding $(B^{(0)} + B^{(1)})^{-1}$ in (3.97) to first order in $B^{(1)}$ gives

$$S_{ji} = \left(\frac{k_j}{k_i}\right)^{1/2}\{(A^{(0)} + A^{(1)})[1 - (B^{(0)})^{-1}B^{(1)}](B^{(0)})^{-1}\}_{ji} \tag{3.98}$$

and extracting the first-order terms we have

$$\begin{aligned} S_{ji}^{(1)} &= \left(\frac{k_j}{k_i}\right)^{1/2}[A^{(1)}(B^{(0)})^{-1} - A^{(0)}(B^{(0)})^{-1}B^{(1)}(B^{(0)})^{-1}]_{ji} \\ &= \left(\frac{k_j}{k_i}\right)^{1/2}[(B^{(0)})^{-1}(B^{(0)}A^{(1)} - A^{(0)}B^{(1)})(B^{(0)})^{-1}]_{ji} \end{aligned} \tag{3.99}$$

Substituting for $A^{(1)}$ and $B^{(1)}$ from (3.86) gives what is called the first-order distorted wave Born approximation[†] to the off-diagonal elements of **S**:

$$S_{ji}^{(1)} = [2i(k_i k_j)^{1/2} B_{ii}^{(0)} B_{jj}^{(0)}]^{-1} \int_{X_0}^{\infty} \psi_i^{(0)}(X) U_{ij}(X) \psi_j^{(0)}(X) \, dX \qquad 3.100$$

This approximation is valid provided that the modulus of all elements $S_{ji}^{(1)}$ are small compared with unity. Unfortunately, for molecular collisions this is frequently not the case because there are large contributions to the integral from the repulsive part of the potential. Another difficulty is that $\mathbf{S}^{(0)} + \mathbf{S}^{(1)}$ is not unitary, so that total transition probabilities are not conserved to unity. We discuss this problem in Section 4.3 and show how unitarity can be imposed.

For the collinear collision of an atom and a diatomic harmonic oscillator under an exponential repulsion, the distorted wave functions can be obtained analytically and are related to Bessel functions. The first-order distorted wave transition probability from the ith to the jth harmonic oscillator level is

$$P_{ji} = |S_{ji}^{(1)}|^2 = \frac{16\pi^2 m^2 (i-j)^2 Y_{ij}^2 \sinh(2\pi k_i/\alpha) \sinh(2\pi k_j/\alpha)}{\alpha^4 [\cosh(2\pi k_i/\alpha) - \cosh(2\pi k_j/\alpha)]^2} \qquad 3.101$$

where Y_{ij} is a matrix element between harmonic oscillator states (3.63):

$$Y_{ij} = \langle \chi_i(Q) | \exp(\alpha Q) | \chi_j(Q) \rangle \qquad 3.102$$

α being the Secrest–Johnson exponential parameter. This result was first obtained by Jackson and Mott[††]. If over the range of the harmonic oscillator functions $\exp(\alpha Q)$ is approximated by $(1 + \alpha Q)$ then Y_{ij} is proportional to $\alpha^{|i-j|}$. Under the normal conditions of scattering theory $2\pi k/\alpha \gg 1$; and for absorption $(i < j)$ the hyperbolic factors in (3.101) can be approximated by

$$\tfrac{1}{4} \exp\left[\frac{\pi(k_i^2 - k_j^2)}{\bar{k}\alpha}\right] \qquad 3.103$$

where $\bar{k}$ is the mean wave vector of the two channels. With these approximations the transition probability (3.92) is proportional to an exponential of the difference in energy of the two channels—the so-called exponential energy gap law. This is the basis of some statistical theories of vibrational energy relaxation.

Thiele and Weare[‡] have compared the distorted wave and exact results for the collinear A + BC collision using the Secrest–Johnson parameterization of the hamiltonian. With the same collision energy and parameters referred to in Table 3.1 they obtained a first-order distorted wave transition probability for the 0–1 transition which was greater than unity, which is unphysical and is due to the non-unitarity of the **S** matrix. Their probability for the 0–2 transition is

[†]The original Born approximation was derived under the assumption that the entire potential could be neglected in deriving the zeroth-order functions; in this case $\psi^{(0)}$ is a plane wave. This approximation is only adequate for high collision energies and potentials without a repulsive core, and therefore finds very little use in atomic and molecular scattering.
[††]J. D. Jackson and N. F. Mott, *Proc. Roy. Soc.*, **A137**, 703 (1932).
[‡]E. Thiele and J. Weare, *J. Chem. Phys.*, **48**, 2324 (1968).

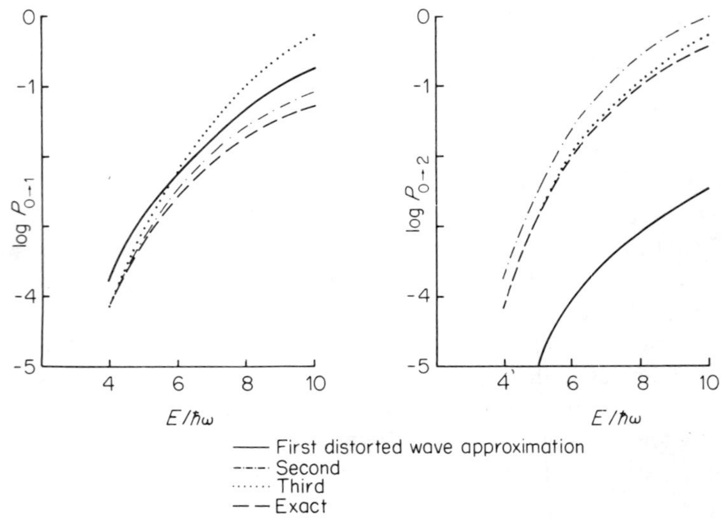

Figure 3.6 Distorted wave transition probabilities for the collinear collision of an atom and a diatomic molecule as a function of total energy. (*Data from E. Thiele and J. Weare,* J. Chem. Phys., **48**, 2324, *1968*)

approximately ten times smaller than the exact result and the second- and third-order results do not improve matters. At lower collision energies the results are better, but only for special values of the parameters do the results appear satisfactory; Figure 3.6 shows some of these. The general unreliability of the distorted wave approximation for vibrational transitions suggests that the energy gap law must be used with some caution.

There is another approach to finding approximate solutions of the multi-channel equations which does not require the off-diagonal coupling terms V_{ij} to be small. It is valid when the collision energy is large compared with the maximum energy that is transferred to the internal states in the collision. If this is the case then in the multichannel equations (3.51) and (3.56) the channel energies E_i^0 or $E_i(r)$ can be replaced by average energies $\bar{E}$ or$\bar{E}(r)$ respectively.

Any orthogonal transformation of the basis will normally lead to off-diagonal terms in the multichannel equations that depend on the differences in the channel energies. However, the channel energy matrix will still be diagonal if all the channel energies are assumed to be equal so that this assumption is a valuable step in decoupling the equations. It is, however, only useful if the transformation matrix is independent of the scattering coordinate; if this is not the case we just move coupling terms from the potential to the kinetic energy terms as in the diabatic to adiabatic transformation.

We shall see in the next chapter that for rotational inelastic scattering there are physically realistic interactive potentials which can be diagonalized by a matrix that is independent of the scattering coordinate, and this leads to various forms of what is known as the sudden approximation. Also, for the A + BC collinear collision the potential energy matrix (3.67) can be factorized into a function of X and a constant matrix if the potential is an exponential function of the distance between the colliding atoms. In this case, which we have already discussed, the sudden approximation is therefore easily made.

Let us write the potential energy matrix (3.67) as

$$V_{ij}(X) = W(X)P_{ij} \qquad 3.104$$

Then $\mathbf{V}$ can be diagonalized by a constant matrix $\mathbf{C}$

$$\mathbf{C}^{\dagger}\mathbf{V}\mathbf{C} = \mathbf{D} \qquad 3.105$$

If E_i^0 can be replaced by an average energy $\bar{E}$ then

$$C^{\dagger}(\mathbf{E}^0 - \mathbf{E} + \mathbf{V})\mathbf{C} \approx \mathbf{C}^{\dagger}(\bar{\mathbf{E}} - \mathbf{E} + \mathbf{V})\mathbf{C} = \bar{\mathbf{E}} - \mathbf{E} + \mathbf{D} \qquad 3.106$$

and as $\mathbf{C}$ is independent of X the kinetic energy operator d^2/dX^2 commutes with $\mathbf{C}$ so that the multichannel equations for the functions defined by

$$\mathbf{\Lambda} = \mathbf{C}^{\dagger}\mathbf{\Psi}\mathbf{C} \qquad 3.107$$

which is a diagonal matrix, are uncoupled and of the form

$$\left(-\mathbf{I}\frac{\hbar^2}{2\mu}\frac{d^2}{dX_2} + \mathbf{I}\bar{E} - \mathbf{I}E + \mathbf{D}\right)\mathbf{\Lambda} = 0 \qquad 3.108$$

where $\mathbf{I}$ is the unit matrix

The $\mathbf{S}$ matrix for the $\mathbf{\Lambda}$ basis is diagonal, as for a distorted wave, and is obtained from the asymptotic behaviour of $\mathbf{\Lambda}$ which is

$$\Lambda \sim \mathbf{I}e^{-i\bar{k}X} - e^{i\bar{k}X}\bar{\mathbf{S}} \qquad 3.109$$

where

$$\bar{k} = \left[\frac{2\mu(E - \bar{E})}{\hbar^2}\right]^{1/2} \qquad 3.110$$

The $\mathbf{S}$ matrix for the basis $\mathbf{\Psi}$ is simply obtained from this by premultiplying by $\mathbf{C}$ and postmultiplying by $\mathbf{C}^{\dagger}$, to give

$$\mathbf{S} = \mathbf{C}\bar{\mathbf{S}}\mathbf{C}^{\dagger} \qquad 3.111$$

If the potential cannot be factorized as (3.104) we can still make progress if the maximum energy transfer in the collision is small compared to the total collision energy. This is equivalent to replacing the oscillator hamiltonian in the Schroedinger equation by a constant E which is the average energy of the relevant oscillator states for the collision—an approximation similar to the introduction of $\bar{E}$ in (3.108). We can then solve the Schrodinger equation

$$\left[-\frac{\hbar^2}{2\mu}\frac{d^2}{dX^2} + \bar{E} - E + V_{A-BC}\left(X - \frac{m_C Y}{m_B + m_C}\right)\right]\Psi(X; Y) = 0 \qquad 3.112$$

for fixed values of Y and obtain solutions whose asymptotic form will be [cf. (3.72)]

$$(X; Y) \sim e^{-i\bar{k}X} - S(Y)e^{i\bar{k}X} \qquad 3.113$$

where $\bar{k}$ is defined by (3.110).

In this approximation, and provided $\Psi(X; Y)$ is a slowly varying function of Y which commutes with the oscillator hamiltonian, a complete wave function is obtained by multiplying (3.113) by the appropriate oscillator wave function:

$$\Psi(X; Y)\chi_i(Y) \qquad 3.114$$

By taking matrix elements of this with $\chi_j(Y)$ and comparing the asymptotic form with (3.72), we obtain **S** matrix elements in the following form:

$$S_{ij} = \int \chi_j(Y)S(Y)\chi_i(Y)\,\mathrm{d}Y \qquad 3.115$$

This is a very general result. If a scattering equation can be solved for fixed values of the internal coordinates **s** and an average channel energy to give a scalar quantity $S(\mathbf{s})$, then an approximate **S** matrix for inelastic collisions is obtained by calculating integrals such as (3.115) over all the internal degrees of freedom, where $\chi_j(\mathbf{s})$ are the channel wave functions.

This approach is generally associated with the description 'sudden approximation', a phrase linked to the idea that the inelastic processes occur in a time that is short compared with the periods of internal motion. This is the justification for calculating S for fixed values of **s**. However, the most important approximation that is made is that the maximum energy transfer to the internal states is small compared to the total collision energy. Thus in a collision of hard molecules the collision time is always very small, but as the energy transfer can be large, as we have seen in the classical treatment discussed in Section 3.1, a sudden approximation will often be rather poor.

Vibrational energy transfer is frequently significant compared with the collision energy, but for diatomic molecules at least, very few quanta of vibration are generally involved. Approximate quantum mechanical solutions are not therefore of great value. For collisions between a light atom and molecules having several vibrational degrees of freedom, the models would become useful. We return to a discussion of sudden approximations in the next chapter because it has been widely applied to rotational inelastic scattering. We then examine its accuracy in more detail.

Chapter 4

Rotationally inelastic collisions

4.1 Classical rotational energy transfer

A molecule will change its rotational energy in a collision if it experiences a torque in a period of time. A torque is produced by a force acting on the molecule which does not pass through the centre of mass and it is measured by the product of the force and the shortest distance between the line of force and the centre of mass. If the torque, Q, acts for a time δt, then there is a change in the angular momentum δJ which we can express as

$$\delta J = \hbar\,\delta j = I\,\delta\omega = Q\,\delta t \qquad 4.1$$

where I is the moment of inertia and $\delta\omega$ the change in the angular velocity. The change in the angular momentum can also be expressed in units of $\hbar$ so as to introduce a rotational quantum number j, which has of course continuous values in classical mechanics.

The change in the rotational energy produced by the torque is

$$\Delta E = \tfrac{1}{2}I(\omega_0 + \delta\omega)^2 - \tfrac{1}{2}I\omega_0^2 \approx I\omega_0\delta\omega \qquad 4.2$$

where ω_0 is the initial angular velocity. This can alternatively be expressed through the change in rotational quantum number

$$\Delta E = \frac{\hbar^2(j_0 + \delta j)^2}{2I} - \frac{\hbar^2 j_0^2}{2I} \approx \frac{\hbar^2 j_0 \delta_j}{I} \qquad 4.3$$

We note that when the moment of inertia is very large a torque which produces a significant change in j may still produce a negligible change in the rotational energy.

The moment of inertia is entirely determined by the atomic masses and the equilibrium geometry. For example, a diatomic molecule of reduced mass μ and equilibrium distance r_0 has $I = \mu r_0^2$. The torque, however, is determined by the intermolecular potential. For neutral systems long-range intermolecular forces are generally weak and in many cases approximately spherically symmetric so they do not give rise to large torques. There are exceptions to this, e.g. when the forces arise from large electrostatic contributions such as dipole–dipole

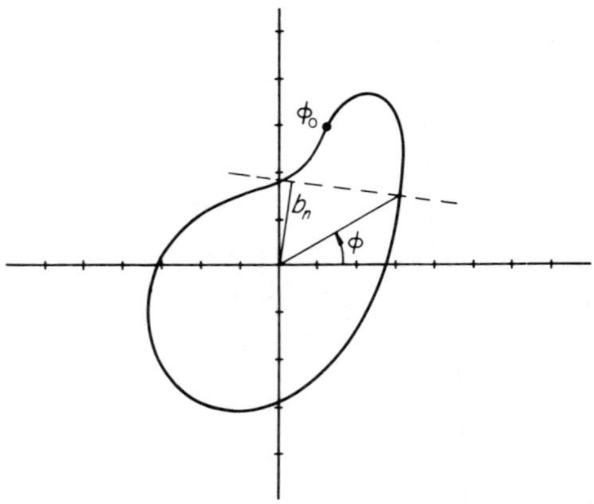

Figure 4.1 Effective impact parameter for a two-dimensional hard potential

forces. In the short range, intermolecular forces are both stronger (repulsive) and generally more anisotropic, so that the torques may be large. Indeed, it is often a good approximation to assume that the changes in rotational state are all due to short-range repulsive forces. If this is the case then the details of the potential are not too important and we can extract the main features by considering the colliding species as hard shapes.

Figure 4.1 shows a two-dimensional hard shape potential that might be appropriate to model the collision of an atom and a polyatomic molecule; the centre of mass of this molecule is at the origin. The line of force on impact is always perpendicular to the contour of the potential so that it is a simple matter to find the perpendicular distance from the centre to the line of force for any point of impact and hence to calculate the torque. This distance which we label b_n is an effective impact parameter for the collision.

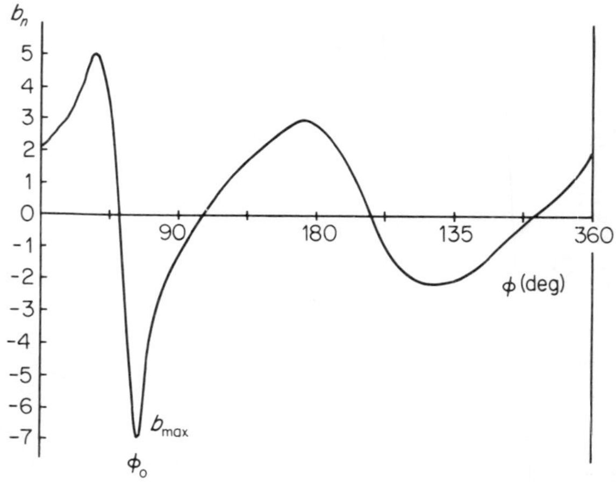

Figure 4.2 Effective impact parameter as a function of orientation angle for the potential of Figure 4.1

The variation in b_n as the point of impact moves around the perimeter is shown for our example in Figure 4.2. The sign of b_n has been defined as positive when the centre of mass is to the left of the line of force and negative when it is to the right. Thus a collision at a point with a positive value of b_n will cause the figure to rotate anticlockwise.

The important feature of Figure 4.2 is that it has maxima and minima. Collisions at points on the perimeter corresponding to these values will produce the greatest changes in the rotational energy, but even more important these will dominate the rotational energy transfer because b_n changes slowly (quadratically) as the point of collision moves away from these optimum points.

The maximum rotational energy transfer will occur for collisions at the point ϕ_0 of the perimeter, shown in Figures 4.1 and 4.2 $[b_n(\phi_0)=b_{max}]$, and if the impact is normal to the perimeter this will produce the maximum torque on the molecule. Such a collision will produce backward scattering, and the change in the angular momentum of the body can be derived from the conservation of the total angular momentum

$$b_{max}(p_i+p_f)=\hbar(j_f-j_i) \tag{4.4}$$

and conservation of energy

$$\frac{p_i^2-p_f^2}{2\mu}=\frac{\hbar^2(j_f^2-j_i^2)}{2I} \tag{4.5}$$

where p_i and p_f are the initial and final values of the relative momentum in the direction of the trajectory, j_f and j_i are the rotational angular momenta and μ is the reduced mass of the system. Eliminating p_f from these equations gives for the maximum change in j of backward-scattered particles

$$\hbar\,\delta j_{max}=2b_{max}\left(p_i-\frac{\hbar j_i\mu b_{max}}{I}\right)(1+\eta)^{-1} \tag{4.6}$$

where

$$\eta=\frac{\mu b_{max}^2}{I} \tag{4.7}$$

The maximum change in the rotational angular momentum for other scattering angles can be obtained in a similar way. The component of p_i tangential to the equipotential curve does not change in a collision but the perpendicular component changes to satisfy the conservation laws (4.4) and (4.5). If χ and χ' are the angles subtended by p_i and p_f with the normal to the surface, as shown in Figure 4.3, then

$$p_i\sin\chi=p_f\sin\chi' \tag{4.8}$$

$$b_{max}(p_i\cos\chi+p_f\cos\chi')=\hbar(j_f-j_i) \tag{4.9}$$

and

$$\frac{p_i^2\cos^2\chi-p_f^2\cos^2\chi'}{2\mu}=\frac{\hbar^2(j_f^2-j_i^2)}{2I} \tag{4.10}$$

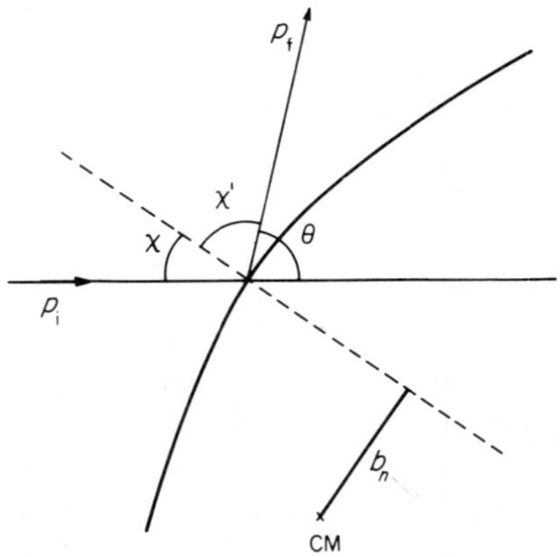

Figure 4.3 Parameters to determine classical energy transfer for a hard shape collision

The scattering angle is

$$\theta = \pi - (\chi + \chi') \qquad 4.11$$

and from these equations we can obtain the maximum change in j as a function of θ and the parameters, p_i, j_i and b_{max}. In the case when the molecule has no initial rotation ($j_i = 0$) the expression is

$$\hbar^2\, \delta j^2(\theta) = 2p_i^2 b_{max}^2 \left[\frac{1 + \eta \sin^2\theta - (1 - \eta^2 \sin^2\theta)^{1/2} \cos\theta}{(1+\eta)^2} \right] \qquad 4.12$$

Determination of the maximum change in j at different scattering angles will therefore give information about b_{max} and if there is more than one peak in the distribution of final j states then this reflects the existence of several non-equivalent turning points on the $b(\phi)$ curve. For example, the potential for the collision of an atom and a homonuclear diatomic molecule in a plane will also give rise to four turning points. These occur as two equivalent pairs and give rise to two peaks in j_f, at $j_i \pm \delta_j$ where δj is given by expressions such as (4.6) or (4.12).

The formulae (4.6) and (4.12) are valid only if multiple collisions are absent. These can occur if the atom transfers most of its energy to the rotation of the molecule at the first impact and it cannot escape from the interaction region before it is hit by another part of the rotating molecule. In such a case the molecule loses rotational energy at the second impact, and clearly the atom will be scattered to any angle. A condition for multiple collision is that $p_f \simeq 0$, and from (4.8) this can only occur for $\chi \simeq 0$, i.e. from head-on collisions. For an initially non-rotating molecule one concludes from equations such as (4.4) and (4.5) that multiple collisions will occur if

$$\frac{\mu b_n^2}{I} \approx 1 \qquad 4.13$$

We now need to see how rotational transitions manifest themselves in the scattering cross-sections, and to do that we must generalize the classical theory of elastic cross-sections given in Section 2.2.

4.2 Classical rotational cross-sections

Cross-sections measure the probabilities of going from initial to final states as a result of the collision, and in classical mechanics we obtain these probabilities by finding the number of classical trajectories that lead from the initial to the final states and emerge at a particular scattering angle.

As in our discussion of elastic scattering cross-sections (Section 2.2), we make the analysis in the centre-of-mass frame, so that a transformation to the laboratory frame is needed, as described in Section 1.5, in order to compare theoretical calculations with experimental cross-sections.

For the collision of two rigid molecules (i.e. no molecular vibrations) the initial state is specified by the initial relative velocity and the values of the components of the rotational angular momenta. Most generally there will be three components of angular momentum for each molecule, but this number will be less for simpler systems such as the collision of an atom and a diatomic molecule. To specify the initial conditions of the trajectory we must, in addition, give the value of the impact parameter, b, and the initial orientations of the molecules. The latter will be specified by a set of angles $\boldsymbol{\alpha}$ whose number is equal to the number of rotational angular momenta components; a maximum of six.

A trajectory that starts with a specific set of initial conditions will terminate with the molecules in particular rotational states, with a particular value of the relative velocity and at a particular scattering angle. However, the final orientation angles are of no significance; indeed as we can stop the trajectory at any time providing the collision is completed (i.e when the intermolecular potential is zero), these angles can have any values. Thus, to obtain the classical cross-sections we must count trajectories that finish in particular final states regardless of the value of these angles.

For atom–molecule or molecule–molecule collisions the scattering pattern does not have the cylindrical symmetry that is present for atom–atom scattering (illustrated in Figure 2.9); i.e. if we change the azimuthal angle of the incoming trajectory, γ, by $\mathrm{d}\gamma$, keeping all other initial conditions constant, the parameters of the final state are *not* unchanged. We must, therefore, calculate the cross-sections by considering trajectories that start in and finish in elemental areas rather than cylindrical areas.

The elemental cross-section area in the incident beam, the analogue of expression (2.20), is

$$\mathrm{d}A = b\,\mathrm{d}b \cdot \mathrm{d}\gamma \cdot \mathrm{d}\boldsymbol{\alpha} \qquad 4.14$$

and this scatters into the solid angle

$$\mathrm{d}\Omega = \sin\theta\,\mathrm{d}\theta \cdot \mathrm{d}\phi \qquad 4.15$$

The differential cross-section for rotationally inelastic scattering from an initial rotational state $\mathbf{j}_i$ to a final state $\mathbf{j}_f$ is now a generalization from the elastic scattering expression (2.22)

$$\sigma_{if}(\Omega)\,\mathrm{d}\mathbf{j}_f = \frac{\text{scattered flux in } \mathbf{j}_f}{\mathrm{d}\Omega} \div \frac{\text{incident flux in } \mathbf{j}_i}{\mathrm{d}A} \tag{4.16}$$

The spread of final rotational states is represented by the factor $\mathrm{d}\mathbf{j}_f$ so that the total scattering from $\mathbf{j}_i$ into the solid angle Ω is

$$\sigma_i(\Omega) = \int \sigma_{if}(\Omega)\,\mathrm{d}\mathbf{j}_f \tag{4.17}$$

The flux of particles is proportional to their velocity; hence we obtain from the above expressions

$$\sigma_{if}(\Omega) = \frac{v_f b\,\mathrm{d}b\cdot\mathrm{d}\gamma\cdot\mathrm{d}\boldsymbol{\alpha}}{v_i\,\mathrm{d}\Omega\cdot\mathrm{d}\mathbf{j}_f} \tag{4.18}$$

As Ω contains two variables (4.15) and the number of components in $\mathbf{j}_f$ is the same as the number of angles in $\boldsymbol{\alpha}$, we see that the number of variables is the same in the numerator and denominator. Moreover, the ratio of these elementary derivatives can be viewed as the transformation between two coordinate systems, and we can write

$$\frac{\mathrm{d}\Omega\cdot\mathrm{d}\mathbf{j}_f}{\mathrm{d}b\cdot\mathrm{d}\gamma\cdot\mathrm{d}\boldsymbol{\alpha}} = \frac{\partial(\Omega, \mathbf{j}_f)}{\partial(b, \gamma, \boldsymbol{\alpha})} \tag{4.19}$$

where the right-hand side is a jacobian determinant defined as

$$\frac{\partial(x_1, x_2, \ldots, x_n)}{\partial(y_1, y_2, \ldots, y_n)} = \begin{vmatrix} \dfrac{\partial x_1}{\partial y_1} & \dfrac{\partial x_1}{\partial y_2} & \cdots & \dfrac{\partial x_1}{\partial y_n} \\ \vdots & & & \vdots \\ \dfrac{\partial x_n}{\partial y_1} & \cdots & \cdots & \dfrac{\partial x_n}{\partial y_n} \end{vmatrix} \tag{4.20}$$

For elastic scattering cross-sections, which were discussed in Section 2.2, the condition $(\mathrm{d}\theta/\mathrm{d}b) = 0$ leads, from (2.23), to an infinite value of the cross-section $\sigma(\Theta)$, and scattering at this angle is called a rainbow. For rotationally inelastic cross-sections we obtain an infinite cross-section when the jacobian (4.19) is zero, which is a more complicated condition. The important point is that the rainbows are still dominant features of the cross-sections.

We can illustrate the characteristic features of inelastic rotational cross-sections by the two-dimensional hard ellipsoid model, a special case of the hard-shape potentials discussed in Section 4.1. This would be a model for the collision between an atom and a homonuclear diatomic molecule. There is only one angular parameter α in this model (see Figure 4.4), which we take as the angle (0 to 2π) between the long axis of the ellipsoid and the z axis, which is the line of the initial velocity in centre-of-mass coordinates. Associated with this

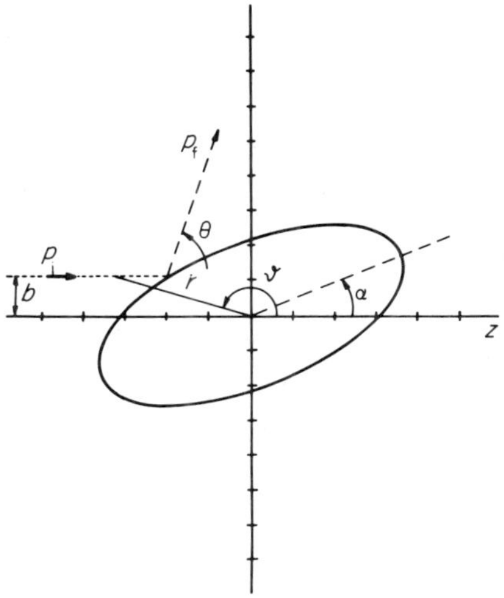

Figure 4.4 Parameters for a hard ellipsoid collision

angle there is a rotational quantum number j which can have positive or negative values corresponding to anticlockwise or clockwise rotation respectively.

In two dimensions the solid angle Ω becomes the scattering angle θ which, as in central field scattering, lies between π and $-\infty$ to allow for orbiting. However, unlike the central field case b can have positive or negative values. The jacobian for this situation is

$$\frac{\partial(\theta, j_f)}{\partial(b, \alpha)} = \begin{vmatrix} \dfrac{\partial \theta}{\partial b} & \dfrac{\partial \theta}{\partial \alpha} \\ \dfrac{\partial j_f}{\partial b} & \dfrac{\partial j_f}{\partial \alpha} \end{vmatrix} \qquad 4.21$$

and the scattering cross-section is, from (4.18),

$$\sigma_{if}(\theta) = \left(\frac{v_f}{v_i}\right)\left[\frac{\partial \theta}{\partial b}\frac{\partial j_f}{\partial \alpha} - \frac{\partial \theta}{\partial \alpha}\frac{\partial j_f}{\partial b}\right]^{-1} \qquad 4.22$$

Figure 4.5 shows the variation of j_f and θ with b for selected values of α, and Figure 4.6, their variation with α for selected values of b.† To simplify the mathematics it has been assumed that the ellipsoid has no initial angular momentum ($j_i = 0$) and that its moment of inertia is very large; the parameters would be appropriate for, say, a collision between He and I_2. The variation $\Delta j = j_f - j_i$ with b and α for other situations would not be substantially different.

If we were to combine in one figure the variation of Δj with both b and α then $\Delta j(b, \alpha)$ would be a surface and a specified value of Δj would be represented

†S. D. Bosanac, *Phys. Rev.*, **A22**, 2617 (1980).

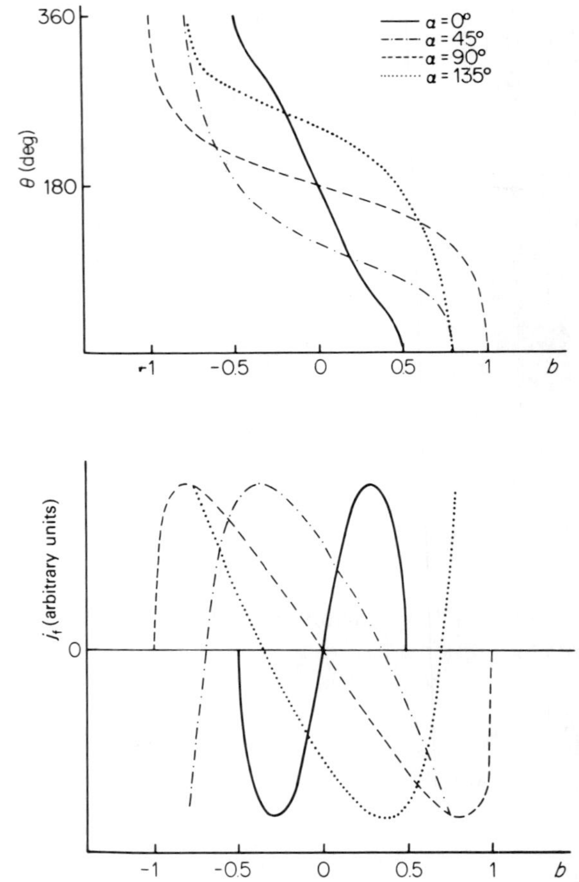

Figure 4.5 Variation of j_f and θ with b ($j_i = 0$) for selected values of α for a hard ellipsoid collision

by a contour of this surface on the (b, α) plane. Likewise a particular value of θ will be represented by a contour of the $\theta(b, \alpha)$ surface. If we were to put both surfaces on the same figure then specific values of both Δj and θ will be given by the intersections of the appropriate contours. Some features of these intersections follow from the principles of topology. For example, if the Δj and θ contours are both closed loops then they may not intersect at all (i.e. there are no trajectories that lead to these values) or they may intersect twice (i.e. two trajectories contribute to the same Δj and θ); more exceptionally they can intersect a higher, even, number of times. Each of these intersections makes a separate contribution to the cross-section so that the total is a sum of terms such as (4.22). In semi-classical theory there would be interference between these trajectories.

If either the Δj or θ surfaces show a maximum or minimum, that is $(\partial(\Delta j)/\partial b)_\alpha$ and $(\partial(\Delta j)/\partial \alpha)_b$ are zero or $(\partial \theta/\partial b)_\alpha$ and $(\partial \theta/\partial \alpha)_b$ are zero, then the jacobian in

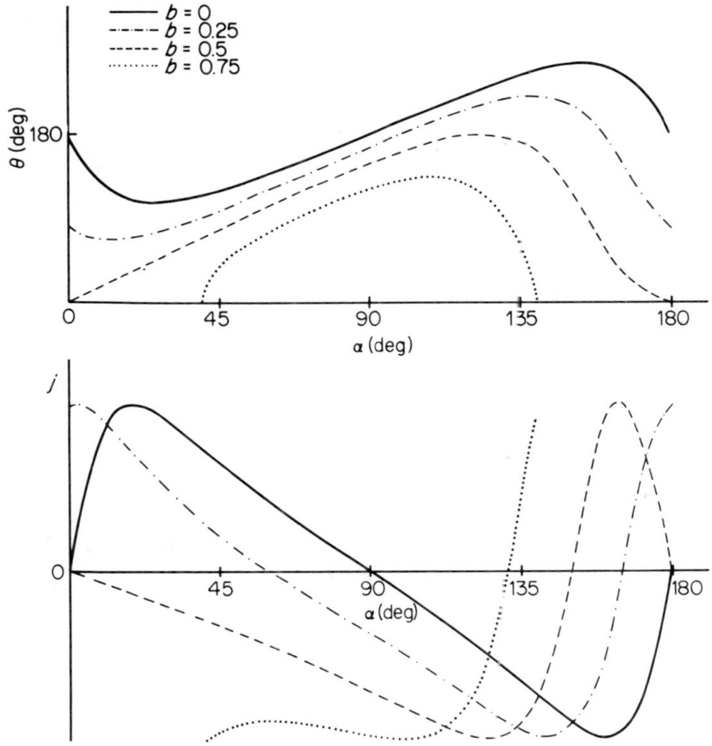

Figure 4.6 Variation of j_f and θ with α ($j_i = 0$) for selected values of b for a hard ellipsoid collision

(4.22) is zero and the cross-section exhibits a singularity. The second of these situations is closely analogous to the rainbows found in elastic scattering and by analogy is called an ordinary rainbow. It generally occurs only if the interaction potential contains attractive and repulsive components. In the first situation, the maximum or minimum in the function $\Delta j(b, \alpha)$ represents the fact that there are certain trajectories that give the largest change in j. Such trajectories are associated with the turning points of the $b_n(\phi)$ graph of Figure 4.2, and we pointed out earlier that the corresponding trajectories dominate the rotational energy transfer.

There is one other topological feature of the $\Delta j(b, \alpha)$ and $\theta(b, \alpha)$ surfaces that is very important and that is when two contours, $\Delta j = \text{constant}$ and $\theta = \text{constant}$, have a common tangent. In this case the jacobian (4.21) is zero, not because any of the partial derivatives are zero but because the two products that occur in (4.22) are equal. The tangent to a contour of constant θ is

$$\frac{db}{d\alpha} = -\frac{(\partial\theta/\partial\alpha)}{(\partial\theta/\partial b)} \qquad 4.23$$

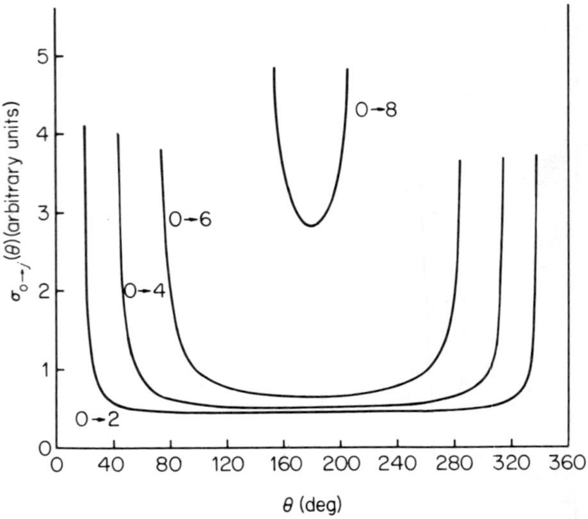

Figure 4.7 Rotational cross-sections for a classical two-dimensional hard ellipsoid collision

and the tangent to a contour of constant Δj is

$$\frac{\mathrm{d}b}{\mathrm{d}\alpha} = -\frac{(\partial \Delta j/\partial \alpha)}{(\partial \Delta j/\partial b)} \tag{4.24}$$

When these are equal the jacobian is zero. This is called a rotational rainbow and it can occur for any value of Δj, not just for the maximum value of Δj.

The cross-section for a transition to a particular j_f has the features shown in Figure 4.7. The scattering angle is shown in the range 0–2π although for two-dimensional scattering and a single collision the pattern is symmetric about $\theta = \pi$. Expression (4.12) gives the maximum change of j as a function of θ, and this occurs at the rainbow angle. In other words, (4.12) gives the rainbow angle for a particular j_f. For an infinite moment of inertia the parameter η, defined by (4.7), is zero, and (4.12) becomes

$$\sin\left(\frac{\theta_R}{2}\right) = \frac{\hbar j_f}{2p_i b_{max}} \tag{4.25}$$

For an elliptic hard shape b_{max} is equal to the difference between the major and minor axes. Expression (4.25) therefore allows us to determine a parameter of the potential by measuring the rainbow angle for a particular j_f.

We have already seen in Figure 4.2 that there can be more than one value of b_{max} and therefore more than one rotational rainbow for a particular j_f. Measurements of these and application of (4.25) to each provides information about the shape of the potential.

If is, of course, possible in hard collisions for all the translational energy to be converted to rotational energy. If this happens then after a short interval of time there will be a further collision which will transfer some energy back to translation. Such collisions lead to asymmetry about $\theta = \pi$. For example, the two rainbow angles in Figure 4.7 will not be symmetrically placed about $\theta = \pi$. Overall it is not possible to convert all the translational energy to rotational energy in hard shape collisions.

We have seen in this chapter that the key quantity for classical rotational scattering is the effective impact parameter b_n and its dependence on the rotational angles. The method for deducing this dependence is given in Appendix 5.

4.3 The scattering matrix

Before we examine the quantum mechanics of rotational inelastic scattering we must look in rather more general terms at some of the quantities that occur in the calculations of inelastic cross-sections. For elastic scattering we have already met the scattering amplitude (2.29) and the phase shift (2.44). For vibrational inelastic scattering we used the **S** matrix (3.72). We now give attention to a more formal examination of scattering matrices.

In an ideal crossed molecular beam experiment we measure the probability of transitions between an initial internal state and a final internal state at each scattering angle, and these probabilities are functions of the collision energy. If we know the intermolecular potential then we can calculate these probabilities. By classical mechanics probabilities are calculated directly, and in quantum mechanics we calculate a scattering amplitude, the probabilities being proportional to the square modulus of the amplitude.

For elastic scattering the cross-sections are determined from the scattering amplitude $f(\theta)$. For inelastic scattering there will be a scattering amplitude for each initial and each final internal state. For example, if we have an incident plane wave with the internal state described by the wave function $\chi_i(\mathbf{s})$, where $\mathbf{s}$ represents all the internal coordinates, then the asymptotic form of the wave function, i.e. the inelastic equivalent of (2.30), is

$$\Psi(\mathbf{r}, \mathbf{s}) \sim \chi_i(\mathbf{s})e^{ik_i z} + \sum_j f_{ji}(\Omega)\chi_j(\mathbf{s})e^{ik_j r}/r \qquad 4.26$$

Collinear scattering, which was dealt with in Section 3.5, is a special case, although the r dependence of the above expression has some similarities to the X dependence of (3.72).

The scattering amplitudes can be represented by a matrix $\mathbf{f}$ with elements that depend on the scattering angle Ω; for scattering without cylindrical symmetry this is a function of the spherical angles θ and ϕ. We showed in Chapter 2 that it was mathematically convenient to represent the elastic scattering amplitude by the partial wave expansion (2.50) in which the phase shifts δ_l appear as angle-independent parameters. We can anticipate that a similar approach will be useful for inelastic scattering amplitudes, which implies that inelastic angle-independent parameters will exist from which the amplitudes can be derived. These parameters are the elements of a matrix, called the scattering or **S** matrix; from it we can predict the state of the system after the collision from a knowledge of the state before the collision.

The internal states of the colliding molecules are defined by wave functions $\chi_j(\mathbf{s})$ which describe collectively their electronic, vibrational and rotational states; these are eigenfunctions of the internal hamiltonian which is a function of the coordinates **s**. The internal state of the system before or after the collision is not necessarily represented by just one of these eigenfunctions but by a wave function

$$\chi(\mathbf{s}) = \sum_j a_j \chi_j(\mathbf{s}) \qquad 4.27$$

where $|a_j|^2$ represents the chance of finding the system in the eigenstate $\chi_j(\mathbf{s})$. Clearly if the eigenstates are specified the state of the system is completely defined by the set of coefficients a_j, which can be written as the column vector

$$\mathbf{A} = \begin{bmatrix} a_1 \\ \vdots \\ a_j \\ \vdots \end{bmatrix} \qquad 4.28$$

The **S** matrix is now formally defined by the equation

$$\mathbf{A}_\mathrm{f} = \mathbf{S}\mathbf{A}_\mathrm{i} \qquad 4.29$$

where $\mathbf{A}_\mathrm{i}$ and $\mathbf{A}_\mathrm{f}$ are the states of the colliding molecules in the asymptotic regions, before and after the collision respectively. This equation defines fully the outcome of the collision if the initial state is fully specified and for this we need to know the details of the experiment. The **S** matrix is a function of the total energy, and for scattering in two or three dimensions depends on the total angular momentum.

The matrix **S** must conserve probability. As $|a_j|^2$ is the probability of being in the jth eigenstate we have

$$\sum_j |a_j|^2 = \tilde{\mathbf{A}}^*\mathbf{A} = 1 \qquad 4.30$$

where $\tilde{\mathbf{A}}^*$ is the adjoint† (transpose and complex conjugate) of **A**. From (4.29),

$$\tilde{\mathbf{A}}_\mathrm{f}^*\mathbf{A}_\mathrm{f} = \tilde{\mathbf{A}}_\mathrm{i}^*\tilde{\mathbf{S}}^*\mathbf{S}\mathbf{A}_\mathrm{i} = 1 \qquad 4.31$$

so that

$$\tilde{\mathbf{S}}^*\mathbf{S} = \mathbf{I} \qquad 4.32$$

or

$$\tilde{\mathbf{S}}^* = \mathbf{S}^{-1} \qquad 4.33$$

This defines **S** as being a unitary matrix. $|S_{ij}|^2$ is the probability that the system makes a transition between the internal states i and j. It can be shown by time

†Also called the hermitian conjugate.

reversal† that this is equal to the probability of the reverse transition j to i and hence **S** must also be a symmetric matrix.

In numerical calculations of the **S** matrix its symmetry and unitarity are important tests of the accuracy of the methods used. As mentioned in Section 3.7 the **S** matrix produced by the distorted wave approximation is not generally unitary but there are a number of algebraic constructions that can convert it into a unitary matrix. The following exponential form is one such:

$$\mathbf{S}^{(0)1/2}\{\exp[(\mathbf{S}^{(0)})^{-1/2}\mathbf{S}^{(1)}(\mathbf{S}^{(0)})^{-1/2}]\}\mathbf{S}^{(0)1/2} \qquad 4.34$$

Expanding the exponential to the second term gives the distorted wave result $\mathbf{S}^{(0)}+\mathbf{S}^{(1)}$.

Let us now suppose that we have reduced the multichannel equations (3.53) to differential equations in one variable, r, the scattering coordinate. For two- or three-dimensional scattering this requires a partial wave expansion of the wave functions $\psi(\mathbf{r})$ but for collinear collisions, discussed in Section 3.5, no angular expansion is necessary. We must now solve these equations subject to the boundary conditions.

If the system is initially in the internal state i, the asymptotic form of this channel wave function can be written

$$\psi_i(r) \sim U_i^-(r) - S_{ii}U_i^+(r) \qquad 4.35$$

where U_i^- and U_i^+ are incoming and outgoing waves respectively. The asymptotic form for another channel j is

$$\psi_j(r) \sim -S_{ji}U_j^+(r), \qquad j \neq i \qquad 4.36$$

which has only an outgoing component. All terms in these expressions are functions of energy and angular momentum.

The incoming and outgoing waves are functions such as

$$U_i^{\pm}(r) = k_i^{-1/2}\exp(\pm \mathrm{i}k_i r) \qquad 4.37$$

where the factor $k_i^{-1/2}$ has been included so that the square modulus of the **S** matrix elements gives fluxes of particles and not probabilities. If there is angular momentum then one may include terms in (4.37) that depend on the orbital quantum number l so as to remove any contributions to **S** that are solely from the centrifugal motion. This is analagous to removing the l-dependent factor from the phase shift in elastic scattering (2.43).

For collinear collisions the transition probabilities are $|S_{ji}|^2$. For three-dimensional or model two-dimensional systems the observables are cross-sections, which have been defined in Section 2.2 for elastic scattering, and these are related to the scattering amplitude by (2.54) and (2.55). We now need to generalize such expressions for inelastic scattering. The generalization requires analysis of the

†The equations of classical mechanics are unchanged when time is reversed and the equations of quantum mechanics are unchanged when time is reversed and Ψ is replaced by Ψ^*. In both cases there is a symmetry in the forward and backward transition probabilities.

coupling of orbital and rotational angular momenta in the collision, and this will be dealt with in the following section. However, anticipating this result we will find that the scattering amplitude which appears in (4.26) is related to the **S** matrix elements by a partial wave expansion to give a result similar to that of (2.51), and which has the general form

$$f_{ij}(\Omega) = \frac{1}{2\mathrm{i}(k_i k_j)^{1/2}} \sum_j g^J_{ji} D^J_{ji}(\Omega)(S^J_{ji} - \delta_{ji}) \qquad 4.38$$

where J is the total angular momentum, g^J_{ji} are coefficients that depend on the degeneracy and other trivial factors but are independent of Ω and the energy, and $D^J_{ji}(\Omega)$ are angular momentum coupling terms, also independent of energy. The cross-section is, by analogy with (2.54),

$$\sigma_{ij}(\Omega) = \left(\frac{k_j}{k_i}\right) |f_{ij}|^2 \qquad 4.39$$

The factor (k_j/k_i) is proportional to the ratio of velocities in the scattered and incident channels and is needed because cross-sections are defined as the ratio of fluxes [see (2.22)]. Cross-sections have the dimensions of area, but in model two-dimensional calculations (scattering in a plane) they will have the dimension of length, and in this case the expression for $f_{ij}(\theta)$ analagous to (4.38) will not have $(k_i)^{1/2}$ in the denominator.

Comparison of (4.38) and (2.51) shows that for elastic scattering in a central potential, for which the only contribution to angular momentum comes from l, the **S** matrix is diagonal with elements related to the phase shift (2.44) by

$$S^l = \exp(2\mathrm{i}\delta_l) \qquad 4.40$$

The term δ_{ji} that appears in (4.38) is the contribution to the cross-section of the unscattered incident plane wave which appears only for the elastic channel. It is usual to simplify (4.38) by defining a transition or **T** matrix by

$$\mathbf{S} = \mathbf{I} - \mathrm{i}\mathbf{T} \qquad 4.41$$

to give

$$f_{ji}(\Omega) = \frac{1}{2(k_i k_j)^{1/2}} \sum_j g^J_{ji} D^J_{ji}(\Omega) T^J_{ji} \qquad 4.42$$

The **T** matrix appears in the asymptotic expressions for the wave function (4.35) and (4.36) as

$$\psi_j(r) \sim \delta_{ji}[U^-_i(r) - U^+_i(r)] + \mathrm{i}T_{ji} U^+_j(r) \qquad 4.43$$

where the initial state is now represented by a standing wave such as $k_i^{-1/2} \sin k_i r$.

Finally, for computational purposes it is often convenient to work with functions that are real standing waves for both the initial and final states as by integrating the multichannel equations the asymptotic limit gives a real symmetric matrix **R**, which is related to the **S** matrix by

$$\mathbf{S} = (\mathbf{I} - \mathrm{i}\mathbf{R})^{-1}(\mathbf{I} + \mathrm{i}\mathbf{R}) \qquad 4.44$$

It should be noted that the literature contains definitions of **T** and **R** (sometimes called **K**) which differ trivially from one another.

As **R** is symmetric, any approximation to **R** that is symmetric will, from (4.44), always give a unitary **S** matrix. This is therefore an alternative to manipulations such as (4.34). In fact the most common approach to the **S** matrix is through approximations to the **R** matrix.

4.4 Quantum theory of rotational inelastic scattering

The classical approach to rotational energy scattering which we have dealt with in Sections 4.1 and 4.2 provides a good qualitative account of average energy transfer and provides state-to-state cross-sections if semi-classical quantization is used. However, it does not give a good description of even moderately resolved state-to-state cross-sections nor of rotational rainbows because both of these are strongly influenced by interference effects. Quantum mechanical calculations on the collisions of heavy atoms and molecules will always be difficult because of the closeness of the energy levels; this is reflected in the large number of quantum states that are usually accessible in the collision. Classical mechanics will normally be adequate for such collisions.

Even for light atoms and molecules a large number of basis functions may be needed to obtain convergence to the solution of the multichannel equations. For a completely general treatment of the collision between two molecules, the basis functions would be products of the ro–vibrational states of both molecules. However, convergence has only recently been achieved for the collision of two non-vibrating diatomic molecules.† Even the simpler problem of the collision between an atom and a diatomic molecule still strains the resources of the largest computer unless the vibrational motion is ignored and the molecule is treated as a rigid rotor—an unsatisfactory model except when the energy is insufficient to open excited vibrational channels.

Rotationally inelastic processes arise from the coupling between the orbital angular momentum of the collision and the rotational angular momentum of the molecules. In three dimensions each rotational state of a molecule is characterized by a set of quantum numbers which defines the total angular momentum and its components along space fixed or molecule fixed axes. In calculating cross-sections one must include all possible coupling of these quantum numbers consistent with the conservation of total angular momentum. There are various ways of making this coupling; the algebra is straightforward but always formidable. However, the difficult task is always in the solution of the multichannel equations. The basic physics of rotational inelastic collisions can be illustrated by the much simpler two-dimensional problem of the collision of atoms and molecules in a plane. In this case all angular momenta are perpendicular to the plane and can therefore be treated as scalar quantities. We

†For example, D. W. Schwenke and D. G. Truhlar, *J. Chem. Phys.*, **88**, 4800 (1988).

now describe the simplest of such problems which is the collision of an atom and a rigid diatomic molecule in a plane.

There are two angles involved in the description of the scattering process, which are defined in Figure 4.4 with respect to space-fixed axes in the centre-of-mass coordinate system. For this problem the hamiltonian in the variables (r, ϑ, α) is

$$\mathbf{H} = -\frac{\hbar^2}{2\mu}\left[\frac{1}{r}\frac{\partial}{\partial r}\left(\frac{r\partial}{\partial r}\right) + \frac{1}{r^2}\frac{\partial^2}{\partial \vartheta^2}\right] - \frac{\hbar^2}{2I}\frac{\partial^2}{\partial \alpha^2} + V(r, \vartheta - \alpha) \qquad 4.45$$

where μ is the reduced mass of the system and I the moment of inertia of the diatomic molecule. The interaction potential depends on r, but only on the relative orientation $(\vartheta - \alpha)$ of the atom and diatomic molecule. The simplest approach is to use the diabatic multichannel equations discussed in Section 3.4, and suitable basis functions are the normalized functions

$$\chi_{lj} = (2\pi)^{-1}\exp(\mathrm{i}l\vartheta + \mathrm{i}j\alpha) \qquad 4.46$$

where l and j, which can be positive or negative integers, are the quantum numbers for orbital and rotational angular momentum respectively. However, this basis does not immediately lead to a separation of the multichannel equations because there will be coupling of terms with different l and j indices.

A better choice is to use what is called a rotating coordinate system with the orientation of the diatomic molecule being defined not to fixed axes in space but to the vector $\mathbf{r}$—a so-called molecule-fixed or body-fixed axis system. The orientation of the diatomic molecule is then defined by the angle

$$\phi = \vartheta - \alpha \qquad 4.47$$

which, from (4.45), is seen to be the angle dependence of the interaction potential.

The basis functions (4.46) can now be written

$$\chi_{Jj} = (2\pi)^{-1}\exp(\mathrm{i}J\vartheta - \mathrm{i}j\phi) \qquad 4.48$$

where $\mathbf{J} = \mathbf{l} + \mathbf{j}$ is the total angular momentum of the system.

To form the multichannel equations we expand each J-conserved solution of the Schrodinger equation in this basis as

$$\Psi(r, \vartheta, \alpha) = \sum_j r^{-1/2} F_{Jj}(r)\chi_{Jj}(\vartheta, \phi) \qquad 4.49$$

Using the hamiltonian (4.45) and taking matrix elements of the basis functions [a step equivalent to that leading from (3.49) to (3.50)] we obtain

$$\left[-\frac{\mathrm{d}^2}{\mathrm{d}r^2} + \frac{(J-j)^2 - \frac{1}{4}}{r^2} + \eta j^2 - k^2 + U_{Jjj}(r)\right]F_{Jj}(r) + \sum_{j' \neq j} U_{Jjj'}(r)F_{Jj'}(r) = 0 \qquad 4.50$$

where

$$\eta = \frac{\mu}{I} \qquad 4.51$$

and

$$E = \frac{\hbar^2 k^2}{2\mu} = \frac{\hbar^2 k_j^2}{2\mu} + \frac{\hbar^2 j^2}{2I} \tag{4.52}$$

The rotational energy of the diatomic molecule is $\hbar^2 j^2/2I$ and k_j is the asymptotic wave vector for channel j. The matrix elements of the interaction potential that appear in (4.50) are

$$U_{Jjj'} = \frac{\mu}{2\pi^2\hbar^2} \int \exp(-\mathrm{i}J\vartheta + \mathrm{i}j\phi) V(r,\phi) \exp(\mathrm{i}J'\vartheta - \mathrm{i}j'\phi)\,\mathrm{d}\vartheta\,\mathrm{d}\phi$$

$$= \frac{\mu\delta_{JJ'}}{\pi\hbar^2} \int V(r,\phi)\exp[\mathrm{i}(j-j')\phi]\,\mathrm{d}\phi \tag{4.53}$$

which are diagonal in J. These matrix elements can be further simplified if the potential is expanded in polar functions

$$V(r,\phi) = \sum_{\lambda=-\infty}^{\infty} K_\lambda(r) \mathrm{e}^{\mathrm{i}\lambda\phi} \tag{4.54}$$

for then

$$U_{Jjj'} = \frac{2\mu}{\hbar^2} \delta_{JJ'} K_{j'-j}(r) \tag{4.55}$$

Equations (4.50) must be solved with the required boundary conditions. The method is formally the same as that used for the collinear A + BC collision; one starts at a value of r for which all $F_{Jj}(r)$ can be taken as zero and with the derivative of one channel function non-zero. Integration outwards produces a function whose asymptotic form can be written in terms of Jost functions (3.70) A^J and B^J. Transformations of these, according to the methods described in Appendix 3, will produce **S** matrix elements, which are defined by imposing an asymptotic form to the solutions of (4.50) as given by (4.35) and (4.36):

$$\Psi_J(r) \sim \delta_{j'j} \mathrm{e}^{-\mathrm{i}(k_j r - l\pi/2)} - S^J_{j'j} \left(\frac{k_j}{k_{j'}}\right)^{1/2} \mathrm{e}^{\mathrm{i}(k_{j'} r - l'\pi/2)} \tag{4.56}$$

where the factors in the incident and scattered waves that depend on the orbital angular momentum, $l = J - j$, have been introduced so that there is no contribution to the **S** matrix elements from the centrifugal potential [the term r^{-2} in (4.50)]. This is equivalent to removing the l-dependent contribution to the phase shift in elastic scattering (2.44).

Unfortunately the initial conditions for integrating the multichannel equations (4.50) are not so easy to establish for rotational scattering as they are for the A + BC collinear collision which was discussed in Section 3.5. The difficulty is that the integration must start from a value of r where $V(r,\phi)$ is very repulsive but where the finite set of basis functions (4.48) is still sufficient to represent the angular dependence of the wave function. If $V(r,\phi)$ is very anisotropic, so that a large number of terms in (4.54) is needed to achieve convergence, then a very

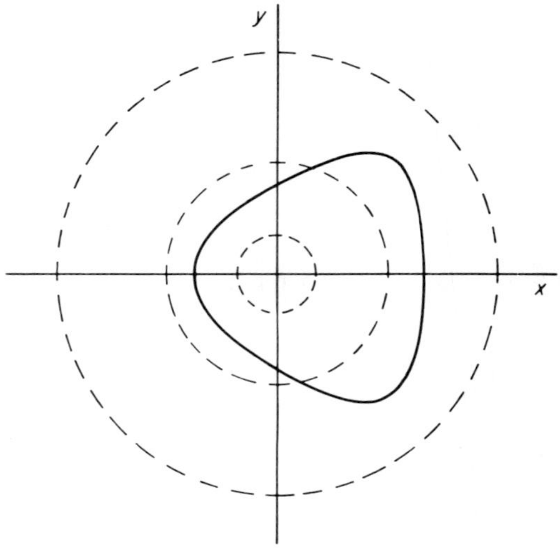

Figure 4.8 Contour of an interaction potential at the classical turning point ($V(\phi) = E$)

large basis may be required to maintain accuracy in the initial stages of integration.

We can see qualitatively that this might be so by examining the contour for the classical turning point

$$V(\phi) = E \tag{4.57}$$

which might be typically as shown in Figure 4.8. The circles r = constant may be wholly outside this contour, and at this distance the rotational basis is rapidly convergent but the radial functions would not be zero. Alternatively, they may be wholly or partly inside where the basis is very slowly convergent (corresponding to a situation in which there is no free rotation of BC against A). The most extreme example of this is for a hard shape collision where there is no value of r from which it is sensible to start the collision except when the shape is circular; in that case there is no rotational inelasticity. There is no satisfactory way of separating the radial and angular wave functions for a general hard shape collision. However, one must not be too pessimistic, because for many atom–diatomic molecule collisions, He–H_2 for example, the anisotropy of $V(r, \phi)$ is not large and the potential is not very 'hard'. In such cases one can find a sensible value of r within the classically forbidden region from which to start the integration.

The asymptotic form of the full scattering wave function will, by analogy with (4.26), be

$$\Psi(r, \theta, \phi) \sim e^{i(k_j z + j\alpha)} + r^{-1/2} \sum_{j'} f_{jj'}(\theta) e^{i(k_{j'} r + j'\alpha)} \tag{4.58}$$

and this can be written as a sum of J-conserved solutions of the form (4.49) where the r-dependence is of the form (4.56). By matching these asymptotic solutions in the same way as for elastic scattering we obtain a partial wave expansion of the scattering amplitude $f_{jj'}(\theta)$. This is a two-dimensional analogue of (4.38) and has the form

$$f_{jj'}(\theta) = \frac{e^{-i\pi/4}}{(2\pi k_{j'})^{1/2}} \sum_{J=-\infty}^{\infty} e^{i(J-j')\theta}(S^J_{j'j} - \delta_{j'j}) \qquad 4.59$$

The cross-section is obtained by inserting this into (4.39).

For the three-dimensional collision of an atom and a rigid diatomic molecule, the hamiltonian will be

$$-\frac{\hbar^2}{2\mu}\nabla^2(\mathbf{r}) + H_{rot}(\alpha, \beta) + V(r, \gamma) \qquad 4.60$$

where the first term is the kinetic energy operator for the relative motion and $H_{rot}(\alpha, \beta)$ is the hamiltonian for the diatomic rotor, which is a function of the azimuthal and polar angles of the intermolecular axis (α and β respectively) in a space-fixed coordinate system. $V(r, \gamma)$ is the interaction potential, where γ is the angle between $\mathbf{r} = (r, \theta, \phi)$ and the diatomic axis, which is related to θ, ϕ, α and β by

$$\cos\gamma = \cos\theta\cos\alpha + \sin\theta\sin\alpha\cos(\phi - \beta) \qquad 4.61$$

The eigenfunctions of H_{rot} are the spherical harmonic functions

$$Y_{jm_j}(\alpha, \beta) \qquad 4.62$$

and its eigenvalues are

$$\frac{\hbar^2 j(j+1)}{2I} \qquad 4.63$$

The eigenfunctions of the angular part of the kinetic energy operator for relative motion are also spherical harmonics

$$Y_{lm}(\theta, \phi) \qquad 4.64$$

Hence the solutions of the Schrodinger equation for the complete system can be written as the expansion

$$\Psi = \frac{1}{r}\sum_{l,m,j,m_j} Y_{lm}(\theta, \phi) Y_{jm_j}(\alpha, \beta) G_{lmjm_j}(r) \qquad 4.65$$

This is a three-dimensional generalization of an expansion in the angular functions of the type (4.49). If we develop multichannel equations for the functions $G_{lmjm_j}(r)$ by taking matrix elements in the space $(\theta, \phi, \alpha, \beta)$ we will in general couple any terms which differ in one or more of the four angular momenta quantum numbers. However, if we form linear combinations of the products of angular functions in (4.65) which are eigenfunctions of the total angular momentum $\mathbf{J}^2$ and its z-component $\mathbf{M}$, then the multichannel equations will decouple into blocks having different values of J or M. The rules for coupling angular momentum functions to give a specified total angular momentum are well known and lead to functions

$$Y_{lj}^{JM} = \sum_{m,m_j} C_{lmjm_j}^{JM} Y_{lm}(\theta, \phi) Y_{jm_j}(\alpha, \beta) \qquad 4.66$$

where $C_{lmjm_j}^{JM}$ are the so-called Clebsch–Gordon coefficients. J can take each of the values

$$l+j, l+j-1, \ldots, |l-j| \qquad 4.67$$

and $M = m + m_j$.

The expansion set

$$\Psi = \frac{1}{r} \sum_{JMlj} Y_{lj}^{JM}(\theta, \phi, \alpha, \beta) G_{JMlj}(r) \qquad 4.68$$

leads to the multichannel equations

$$\left[\frac{\mathrm{d}^2}{\mathrm{d}r^2} - \frac{l(l+1)}{r^2} + k_j^2\right] G_{JMlj}(r) = \frac{2\mu}{\hbar^2} \sum_{l'j'} V_{ljl'j'}^{JM} G_{JMl'j'}(r) \qquad 4.69$$

which are solved using the standard procedure described for two-dimensional collisions. By expanding the potential in angular functions

$$V(r, \gamma) = \sum_{\lambda=0}^{\infty} K_\lambda(r) P_\lambda(\cos \gamma) \qquad 4.70$$

where $P_\lambda(\cos \gamma)$ are Legendre polynomials, which is a set of orthogonal functions. Only even λ contribute to the potential for a homonuclear diatomic in three-dimensional space. The matrix elements of V that appear in (4.69) then have the form

$$V_{ljl'j'}^{JM} = \sum_{\lambda} K_\lambda(r) \langle jl; J | P_\lambda(\cos) | j'l'; J \rangle \qquad 4.71$$

where the matrix element of P_λ is non-zero only if λ satisfies coupling rules similar to (4.67), but with only alternate terms:

$$\lambda = \begin{cases} j+j', j+j'-2, \ldots, |j-j'|+1 \text{ or } |j-j'| \\ l+l', l+l'-2, \ldots, |l-l'|+1 \text{ or } |l-l'| \end{cases} \qquad 4.72$$

Explicit expressions for these matrix elements have been given by Brink and Satchler.†

We stress that, as for the two-dimensional case (4.54), the expansion (4.70) may be very slowly convergent, particularly in the repulsive region of the potential. For example, Figure 4.9a shows the $V = 0$ contour for the collision of He with an excited state of Li_2 which is known to be very anisotropic. Figure 4.9b shows the result of fitting just two terms ($\lambda = 0, 2$) to *ab initio* data on the collinear and perpendicular (T-shaped) structures.

For the collision of an atom and a homonuclear diatomic molecule (or other

†D. M. Brink and G. R. Satchler, *Angular Momentum*, 2nd ed., Oxford University Press, 1968.

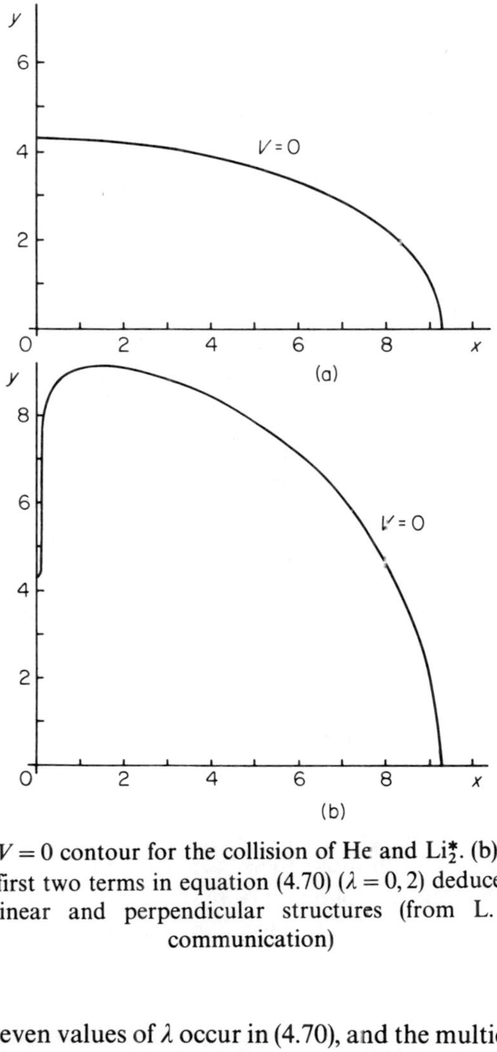

Figure 4.9 (a) The $V = 0$ contour for the collision of He and Li_2^*. (b) The $V = 0$ contour obtained from the first two terms in equation (4.70) ($\lambda = 0, 2$) deduced by fitting only to data for the collinear and perpendicular structures (from L. Cooper, personal communication)

$D_{\infty h}$ system) only even values of λ occur in (4.70), and the multichannel equations for each J separate into two blocks, one for even j and one for odd j. It follows that only $\Delta j = 2$ transitions are allowed.

From the solution of the multichannel equations one obtains S-matrix elements for each value of J, and the scattering amplitude is then given by the following expression due to Arthurs and Dalgarno.†

$$f(jm_j j'm_{j'}; \theta, \phi) = \sum_J \sum_l \sum_{l'} \sum_{m'} \left[\frac{(2l+1)\pi}{k_j k_j'} \right]^{1/2} i^{l-l'+1} C^{Jm_j}_{l0jm_j} C^{Jm_j}_{l'm'j'm_{j'}} \times [S^J_{j'l'jl} - \delta_{ll'}\delta_{jj'}] Y_{l'm'}(\theta, \phi) \qquad 4.73$$

†A. M. Arthurs and A. Dalgarno, *Proc. Roy. Soc.*, **A256**, 50 (1960).

The differential cross-section is, by analogy with (4.39),

$$\sigma(jm_j j' m_{j'}; \theta, \phi) = \left(\frac{k_{j'}}{k_j}\right) |f(jm_j j' m'_j; \theta, \phi)|^2 \qquad 4.74$$

and the total cross-section is the integral of this over θ and ϕ, which, as for elastic scattering (2.57), simplifies because of the orthogonality of the functions $Y_{lm}(\theta, \phi)$:

$$\sigma(jj') = \pi(2j+1)^{-1} k_j^{-2} \sum_{J=0}^{\infty} \sum_{l} \sum_{l'} (2J+1) |S^J_{jlj'l'} - \delta_{jj'}\delta_{ll'}|^2 \qquad 4.75$$

Unless molecules are oriented in the incident and scattered beams, the measured cross-sections will be an average of initial m_j and a sum over final $m_{j'}$ states:

$$\sigma(jj'; \theta, \phi) = (2j+1)^{-1} \sum_{m_j} \sum_{m_{j'}} \sigma(jm_j j' m_{j'}; \theta, \phi) \qquad 4.76$$

The same average for the total cross-section gives a particularly simple expression, which was derived by Arthurs and Dalgarno.

The quantum numbers m_j and $m_{j'}$ that occur in (4.66) and (4.73) define the components of the diatomic angular momentum in a space-fixed axis system, usually defined with respect to the incident beam. It is more convenient to measure orientation with respect to the linear momentum of the particles at all times, which is called the helicity representation. This means that m_j and the helicity are the same in the incident beam but they are different in the scattered beam; the helicity eigenfunctions will be a linear combination of $m_{j'}$ states, and this is equivalent to a transformation from a space-fixed to a body-fixed coordinate system analagous to that discussed for two-dimensional scattering (4.47).

The multichannel equations in the helicity representation are simpler in their angular components than those in the space-fixed coordinate system although there are now off-diagonal coupling terms in the centrifugal potential r^{-2} [which is diagonal in (4.69)]. The multichannel equations and expressions for the scattering amplitudes and cross-sections in this representation were derived by Pack and Hirschfelder.†

4.5 Approximate schemes for rotational inelasticity

In Section 3.7 we discussed approximate schemes for solving the coupled channel equations for collinear atom–diatomic molecule collisions. However, we noted that with current computational power they were not needed except to provide physical insight into the collision process. For rotational scattering, or even more when rotations and vibrations are simultaneously excited, there is a much greater need for approximate solutions because the number of coupled channel equations necessary to obtain convergence in the **S** matrix may be very large.

†R. T. Pack and J. O. Hirschfelder, *J. Chem. Phys.*, **49**, 4009 (1968).

The distorted wave Born approximation (3.100) has been used for atom–rigid diatomic molecule collisions, and if the unitarity condition is imposed, using (4.34), the results are quite satisfactory for low-energy collisions and small ellipticity of the intermolecular potential.†

Much the more common approach for simplifying the coupled channel equations has been to use what are called sudden approximations in one of several forms; the key expressions are 'energy sudden', 'centrifugal sudden' and 'infinite-order sudden' (IOS).

A diagonalization of the potential coupling terms, as in (3.105), when applied to (4.69) will lead to off-diagonal terms in $l(l+1)/r^2$ and k_j^2. In the centrifugal sudden approximation one assumes degeneracy of the l quantum numbers before diagonalization, replacing $l(l+1)$ by an average $\overline{l(l+1)}$ for all relevant orbital angular momenta in the coupling procedure. The first benefit of this is that a transformation of the basis can be found in which there is no coupling between terms of different orbital angular momentum. The appropriate basis is in fact the set of rotational functions in a body-fixed axis system. These equations, which are still coupled in the potential energy terms through j, are called the coupled-state equations.

It has been argued that this approximation is valid for repulsive potentials where there is only one classical turning point for a given l (see Figure 2.2), and this turning point does not change rapidly with l; in effect it is being assumed that the centrifugal potential is not the dominant feature of the interaction. For interaction potentials with an attractive well the approximation is questionable. Obviously if j changes in a collision then l must change also, to conserve total J, and the question is whether the change in the centrifugal potential is important or not.

If we adopt the coupled state approximation and then diagonalize the potential (this can be done by a transformation matrix which is independent of r if there is only one non-spherical term in the Legendre expansion of the potential) then we will introduce off-diagonal terms in k_j^2. If these are ignored we arrive at what is called the centrifugal sudden approximation. The centrifugal potential $l(l+1)/r^2$ is still diagonal and the terms in k_j^2 are still treated as diagonal, but not equal.

In the energy sudden approximation one replaces k_j^2 by an average channel energy $\bar{k}^2$, so that no off-diagonal terms in k_j^2 will be produced by any transformation of the basis. This will produce off-diagonal terms in the centrifugal potential, but again these are ignored. In either of these sudden approximations the coupled channel equations are completely decoupled and solved independently, the final **S** matrix being produced by an inverse transformation such as (3.111).

If the interaction potential cannot be factorized into a radial part and an angular part, and this is generally true for very anisotropic potentials, then there is no diagonalization matrix independent of r, so that the energy sudden and

†S. D. Bosanac and G. G. Balint–Kurti, *Molec. Phys.*, **29**, 1797 (1975).

centrifugal sudden approximations are useful in a very limited number of cases.

The IOS approximation was originally derived by introducing average k and average l in the coupled channel equations. However, this again would only be useful if the interaction potential can be factorized. Alternatively this approximation can be placed in the same family as the constant energy approximation already described for atom–diatom collinear collisions, and expressed by equation (3.112). The **S** matrix elements are obtained for fixed relative orientations of the colliding molecules (for atom–diatom collisions there is only one angular variable γ). Such a calculation does not necessarily require a Legendre expansion of the potential; indeed there is nothing to be gained by making such on expansion. The full **S** matrix is then calculated, as in (3.115), by averaging over the orientation angles.

The IOS approximation can be derived from the coupled channels equations (4.69) by multiplying by the angular functions (4.66), summing over all indices and using the so-called closure rule for complete sets of states to obtain separable equations in which the interaction potential appears as a function of the internal angles of the collision. However, this approach is rather artificial; firstly, we expand the wave function in an angular basis and then we use the closure rule to return to functions in which the angles are parameters. The direct approach is the one we followed for collinear collisions, which is to solve the Schrodinger equation for fixed values of the internal coordinates. For this procedure we have a choice of space-fixed or body-fixed axis systems.

It is more natural to work in the body-fixed system because the interaction potential depends on γ and this angle is one of the independent variables of the angular momentum operators—for the rotation of the diatomic molecule about the body-fixed z axis.

The hamiltonian in the body-fixed system is†

$$-\frac{\hbar^2}{2\mu r}\frac{\partial^2}{\partial r^2}r+\frac{\mathbf{j}^2}{2I}+\frac{(\mathbf{J}-\mathbf{j})^2}{2\mu r^2}+V(r,\gamma) \qquad 4.77$$

The diatomic angular momentum **j** is a function of γ and an arbitrary azimuthal angle (ε say) which the diatomic axis makes with the body-fixed x axis. The total angular momentum **J** is a function of the angles (θ, ϕ) that the body-fixed z axis makes with respect to space-fixed axes. If we neglect **j** in the centrifugal term then there is no coupling of the rotational states by the angular momentum terms; this is equivalent to the coupled states approximation. If we go further and make the energy sudden approximation, replacing $\mathbf{j}^2/2I$ by an average rotational energy $\bar{E}$, as in (3.112), then the Schrodinger equation is

$$\left[-\frac{\hbar^2}{2\mu r}\frac{\partial^2}{\partial r^2}r+\bar{E}+\frac{\mathbf{J}^2}{2\mu r^2}+V(r,\gamma)-E\right]\Psi(r,\theta,\phi;\gamma)=0 \qquad 4.78$$

†C. F. Curtis was the first to formulate the scattering equations in the body-fixed system and there have been many descriptions of this procedure. A good account of the relative merits of space-fixed and body-fixed systems can be found in R. T. Pack, *J. Chem. Phys.*, **60**, 633 (1974).

which is a single-channel equation parametrically dependent on γ. A complete wave function is obtained by multiplying Ψ by the appropriate rotor wave function which is $Y_{jm}(\gamma, \varepsilon)$. The wave function $\Psi(r, \theta, \phi; \gamma)$ is subject to the scattering boundary condition

$$\Psi(r, \theta, \phi; \gamma) \simeq e^{i\bar{k}z} + \frac{e^{i\bar{k}r}}{r} f(\theta, \phi; \gamma) \qquad 4.79$$

where $\bar{k} = 2\mu(E - \bar{E})/\hbar^2$. By the logic of the derivation of (3.115), we obtain the inelastic scattering amplitude for the transition $jm_j \rightarrow j'm_{j'}$:

$$f(jm_j j'm_{j'}; \theta, \phi) = \int Y^*_{jm_j}(\gamma, \varepsilon) f(\theta, \phi; \gamma) Y_{j'm'_j}(\gamma, \varepsilon) \sin\gamma \, d\gamma \, d\varepsilon \qquad 4.80$$

which is known as the IOS scattering amplitude.

As we are working in a body-fixed axis system m_j and $m_{j'}$ do not refer to the same space-fixed z axis and a further transformation is needed to go into the laboratory frame. The details of this can be found in the article by Pack just referred to.

In solving equation (4.80) we are essentially dealing with an elastic scattering problem for each angle γ and we can use the same approximations as for the central force problem described in Chapter 2. In particular, the stationary phase approximation (2.66) and the JWKB wave functions (Section 2.4) can be used to greatly simplify the computing requirement in the IOS method, as well as giving a better insight into the inelastic scattering problem.

Figure 4.10 shows partial cross-sections

$$\sigma^J(jj') = (2J + 1)(2j + 1)^{-1} \sum_{ll'} |S^J_{jlj'l'}|^2 \qquad 4.81$$

calculated for two transitions using a model potential (4.70):

$$V(r, \gamma) = K_0(r) + K_2(r) P_2(\cos\gamma) \qquad 4.82$$

with K_0 and K_2 being Lennard–Jones 6–12 functions parameterized for $Ar + N_2$ collisions.† The sum of $\sigma^J(jj')$ is proportional to the m-averaged total cross-section (4.76).

The partial cross-sections show oscillations with J which are slow in this example. For a very anisotropic potential the oscillations will be much more rapid. The orbital quantum number l, which is proportional to the classical impact parameter b, runs from $|J - j|$ to $J + j$, so that when j is small, large J can be connected to collisions having large impact parameters. It is for this reason that σ^J tends to zero at large J.

The results of Figure 4.10 are for the most probable collision energy in a gas at 300 K. Up to 36 channels (l values) were required to converge the close coupled calculations; more channels are required at low J than at high J. The IOS results were obtained using the JWKB approximation for the phase shift

†T. P. Tsien, G. Parker and R. T. Pack, *J. Chem. Phys.*, **59**, 5373 (1973).

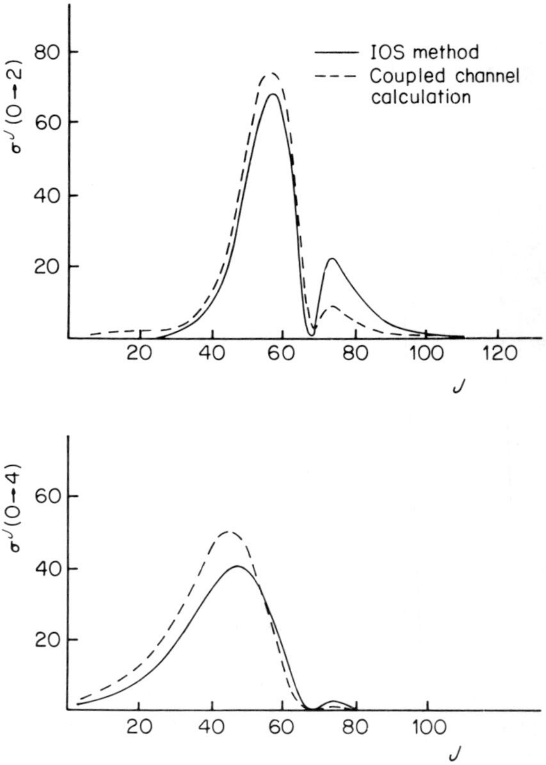

Figure 4.10 Partial cross-sections for a potential (4.82) with parameters appropriate to the $Ar + N_2$ collision

(2.80). There is qualitative agreement between the close coupled and IOS results but the total cross-sections obtained from the IOS method are 18 per cent too high for $0 \rightarrow 2$ and 77 per cent too high for $0 \rightarrow 4$. Many studies, however, have shown that the IOS method becomes poorer as the potential becomes more anisotropic.

If we measure the anisotropy of the potential by the value of $r(\theta)$ at which the potential changes from being attractive to being repulsive, then in the above example the major axis is $r(0) = a = 3.67$ Å and the minor is $r(\pi/2) = b = 3.37$ A°. Approximating the surface $V(\theta) = 0$ by an ellipsoid one can show that b_{max}, as used in equation (4.4), is equal to $a - b$, which is 0.30 Å. The maximum change in j occurs for backward collisions ($\theta = \pi$) and from (4.12) we deduce a classical expression for the maximum change of j at this angle which is

$$\delta j_{max} = \frac{2p_i b_{max}}{\hbar(1 + \eta)} = \frac{2k_i(a - b)}{1 + \eta} \qquad 4.83$$

where η, given by (4.7), is equal to 0.18 for the $Ar{-}N_2$ collision. For a room temperature collision $k_i = 14$ Å^{-1}, which gives $\delta j_{max} = 7$. At this collision energy the transition $j = 0$ to $j = 10$ is open but the classical picture shows that only part of the collision energy can be transferred. No coupled-channel calculations have been reported for transitions above 0–4 and we would expect 0–6 to be weak and 0–8 probably not detectable. The IOS method is valid only for small energy transfer and would not be expected to be accurate for these high Δj transitions.

The semi-classical trajectory method (classical **S** matrix), which was described in Section 3.6, has also been used to calculate inelastic transition probabilities for the collision of an atom and a diatomic rigid rotor. We remind the reader that the method involves finding trajectories which go from specified initial to specified final states and calculating the action integral along these trajectories. For the A + BC collisions these states are specified by the values of j and l, both selected by semi-classical quantization. The transformation from collision to action-angle variables for this system has been given by Miller.[†]

Miller calculated classical S-matrix elements for a potential of the form

$$V(r, \gamma) = K(r)[1 + P_2(\cos \gamma)] \qquad 4.84$$

$K(r)$ being a Lennard–Jones function.[††] Results were compared with close-coupled values for $J = 6$. There was good qualitive agreement, with errors in the partial cross-sections typically of the order of 10 per cent.

The accurate calculation of rotational cross-sections is a formidable task; only for atom–diatomic molecule collisions and potentials which have a weak anisotropy is the problem computationally straightforward. The scale of the problem increases substantially for the collision of two diatomic molecules and for the collision of an atom and a polyatomic molecule. Potentials for such systems are very uncertain and large coupled-channel calculations have to be solved. Nevertheless, experimental studies of rotational cross-sections show that they often conform to simple fitting or scaling laws. The theoretical justification for such laws may not always be well established but can be very valuable in analysing experimental results. In particular, if a small number of cross-sections are known experimentally then these can be used to determine the parameters in an empirical law and the complete set of cross-sections can then be predicted.

The simplest and earliest of the empirical laws to be proposed was the exponential gap law. In this the probability for rotational energy transfer decreases exponentially with the ratio of the energy transferred to the kinetic energy of the collision. This is analagous to the exponential gap law for vibrational energy transfer which was mentioned in Section 3.6.

Polanyi and Woodall[‡] analysed the distributions of the rotational states of HCl produced by the reaction

$$H + Cl_2 \rightarrow HCl + Cl$$

by monitoring the infrared emission spectra. The rotational distributions deviate

[†]W. H. Miller, *J. Chem. Phys.*, **54**, 5386 (1971).

[††]Note that $V = 0$ when $K(r) = 0$, so that the contour surface for this value of the potential is spherical; overall the potential is not very anisotropic.

[‡]J. C. Polanyi and K. B. Woodall, *J. Chem. Phys.*, **56**, 1563 (1972).

strongly from an equilibrium Boltzmann distribution and when there is a low pressure in the reaction chamber the distributions peak at high values of j. With increasing pressure these distributions are relaxed by collisions towards the Boltzmann distribution, and by following the pressure dependence of the emission spectra Polanyi and Woodall were able to reach some conclusions about the laws governing rotationally inelastic collisions. They concluded that the exponential gap law was superior to three other models which they examined; these were that the probabilities of transitions $\Delta j = \pm 1$ were proportional to, independent of or inversely proportional to j. They later concluded that the data could not distinguish between single or multiple quanta (up to $\Delta j = \pm 5$) transitions.

Their most important finding was that if there were an initial peak in the distribution at high j then this peak did not shift during the relaxation but just became less prominent. In contrast, from the model in which the transition probability was proportional to j the peak gradually moved to lower j and finally became the peak in the Boltzmann distribution.

The exponential gap law was later rationalized by Levine and coworkers using statistical arguments and information theory analysis; this topic will be described in Section 5.5. In the currently used form of the exponential gap law we write the rate constant for the transition from j to j' (or alternatively the cross-section for this process) as

$$k_{jj'} = Ck^0_{jj'} \exp\left(-\frac{\theta_{\mathbf{R}}|\Delta E_{jj'}|}{E} \right) \qquad 4.85$$

where $\Delta E_{jj'}$ is the energy transferred and E is the energy available in the collision for redistribution; C and $\theta_{\mathbf{R}}$ are constants, independent of j and j' but possibly depending on E; $k_{jj'}$ is the so-called prior value for k which takes account of the density of available states at level j'. This is proportional to the degeneracy of the level, $2j + 1$, and to the density of states in the translational degree of freedom, which is proportional to $(E - E_{j'})^{1/2}$ [see (5.45)]. The function $\ln(k_{jj'}/k^0_{jj'})$ is called the surprisal, and a linear surprisal [see (5.51)] is consistent with (4.85).

Whilst the exponential gap law holds quite well, particularly for light diatomic molecules, it is less satisfactory for very anisotropic systems and also for large energy transfer. Pritchard and coworkers found that a power gap law [replacing the exponential function in (4.85) by $|\Delta E_{jj'}|^{-\alpha}$] was better for the rotational relaxation of Na_2^* with a number of different colliding molecules. Further improvement was found (not surprisingly) by combining exponential and power behaviour in the expression

$$k_{jj'} = k^0_{jj'} \left| \frac{\Delta E_{jj'}}{B} \right|^{-\alpha} \exp\left(-\frac{\theta_{\mathbf{R}}|\Delta E_{jj'}|}{E} \right) \qquad 4.86$$

where B is the rotational constant of the diatomic molecule (in energy units). Optimum values of α vary between about 0.9 (for I_2^*–Xe) and 1.6 (for Li_2^*–Li).[†]

[†]T. A. Brunner and D. E. Pritchard, Fitting laws for rotationally inelastic collisions, in K. P. Lawley (ed.), *Dyanmics of the Excited State*, John wiley and Sons, 1982.

The power gap laws are not always successful. For example, in the collisional excitation of the lowest rotational state of NO by Ar an odd–even oscillation in the quantum number of the final-state cross-sections has been observed.† These observations were found to be consistent with coupled-channel calculations using model potentials, and it appears that the oscillations derive from strong anisotropy involving both categories. NO, for example, has a small (but not zero) dipole moment and a large quadrupole moment, and the scattering appears to be strongly influenced by both P_1 and P_2 terms in the potential. A number of other fitting laws have been proposed to describe these difficult cases but one probably should conclude that, with complicated potentials, one will get complicated cross-sections; there will be some collisions that fall into no simple pattern.

†P. Andresen, H. Joswig, H. Pauly and R. Schinke, *J. Chem. Phys.*, **77**, 2204 (1977).

Chapter 5

Reactive scattering

5.1 Coordinates for reactive collisions

We can think of reactive scattering as a category of inelastic scattering in which the internal states depend not only on the rotational, vibrational and electronic energies of the fragments but also on the manner in which the atoms are grouped together in the initial and final states. For example, in the simple atom–diatomic molecule exchange reaction represented symbolically as

$$A + BC \rightarrow AB + C \qquad 5.1$$

the internal states will be both BC states and AB states, and if there is also the possibility of forming AC molecules or of having sufficient energy in the reaction to give A + B + C, then these must be added to the list of internal states.

The quantum mechanical treatment of inelastic scattering, which we described in Chapters 3 and 4, relies for its success on finding good basis functions for the internal states. A prerequisite for this is that the coordinates can be separated into an internal set $\mathbf{s}$ and scattering coordinates $\mathbf{r}$. It is much more difficult to make such a separation when the internal states encompass reaction than when they do not. For example, in scheme (5.1) the internal coordinates for the reactants would be the B—C bond length and the angles which the BC axis subtends in a centre-of-mass coordinate system. For the products, however, the internal coordinates will be the A—B bond length and the angles of the AB axis. Thus, not only will the wave functions of the internal states evolve during the scattering process but the coordinates on which these wave functions are based must also evolve. Because of this the quantum mechanical treatment of reactive scattering is much more difficult than that of inelastic non-reactive scattering, and it is for this reason that we treat reactive scattering as a separate category.

There are two approaches to the coordinate problem. One of these is to use different coordinates in different regions of space, and in particular to use different coordinates for reactants and for products. In this case, the basis functions which describe the internal states must change with the coordinates and it will be necessary to maintain continuity of the wave functions at the boundaries. The second method, to be described later, is to use coordinates which

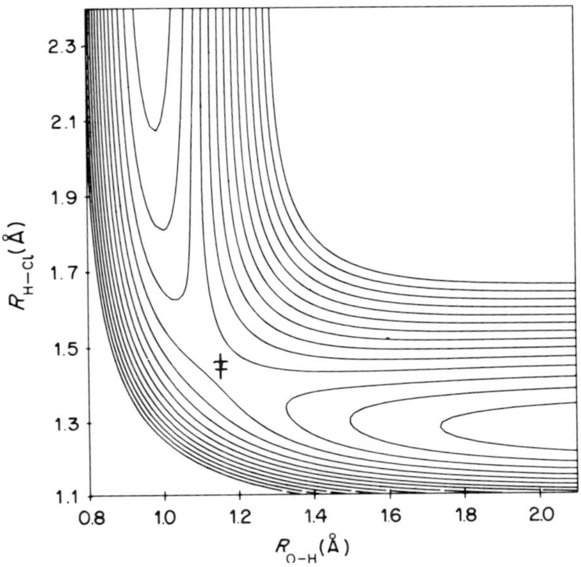

Figure 5.1 Contours of a LEPS potential for the collinear reaction

$$O(^3P) + HCl(^1\Sigma^+) \rightarrow OH(^2\Pi) + Cl(^2P)$$

Lowest contour – 104 kcal/mol, contour spacing 2 kcal/mol. (*Parameters from A. Persky and M. Broida,* J. Chem. Phys., **81**, *4352, 1984*)

cover the whole space but which are not a linear transformation of either the reactant or product coordinate sets. In this case the best coordinates may depend on the specific problem to be solved. For reaction (5.1), for example, coordinates that are suitable for the exchange of light atoms A and C with a heavy atom B may be unsuitable if A and C are heavy and B is light.

The only reactive system for which a relatively simple separation into internal and external coordinates can be made is the collinear atom–diatomic molecule reaction—reaction (5.1) when the three atoms are constrained to a line. Although this model cannot give cross-sections, only reaction probabilities, it is relevant to several important reactions where it is known that the easiest route to reaction is along a collinear path.

Figure 5.1 shows contours of the ground-state potential for the reaction

$$O(^3P) + HCl(^1\Sigma^+) \rightarrow OH(^2\Pi) + Cl(^2P) \qquad 5.2$$

along the collinear path. This reaction is almost thermoneutral and the saddle point is almost symmetrically placed between the entrance and exit valleys.

The coordinates used to represent the potential in Figure 5.1 are symmetrical with respect to reactants and products; one is the reactant diatomic bond length R_{BC} and the other the product diatomic bond length R_{AB}. The classical kinetic energy in this coordinate system is

$$T = \frac{1}{2M}[m_A(m_B + m_C)\dot{R}_{AB}^2 + m_C(m_A + m_B)\dot{R}_{BC}^2 + 2m_A m_C \dot{R}_{AB}\dot{R}_{BC}] \qquad 5.3$$

where

$$M = m_A + m_B + m_C \qquad 5.4$$

For the analysis of the collinear A + BC non-reactive collisions given in Chapter 3 we used coordinates in which the kinetic energy was separable into the kinetic energy of the BC vibration and the kinetic energy for the relative motion of A to the centre of mass of BC. We see from (5.3) that by using both diatomic relative velocities as variables we have lost this separability. This means that even if the potential energy is separable in these coordinates, say,

$$V = V_{AB}(R_{AB}) + V_{BC}(R_{BC}) \qquad 5.5$$

there will be coupling from the cross terms in the kinetic energy. As a result the atomic masses play an important role in the dynamics of such a reaction.

To show the importance of the atomic masses it is common practice to make a further transformation of variables which is equivalent to plotting the potential in a skew coordinate system, shown in Figure 5.2. The rectangular coordinates, (X, Y) in this construction, are related to R_{AB} and R_{BC} by

$$X = aR_{AB} + bR_{BC}\cos\beta$$
$$Y = bR_{BC}\sin\beta \qquad 5.6$$

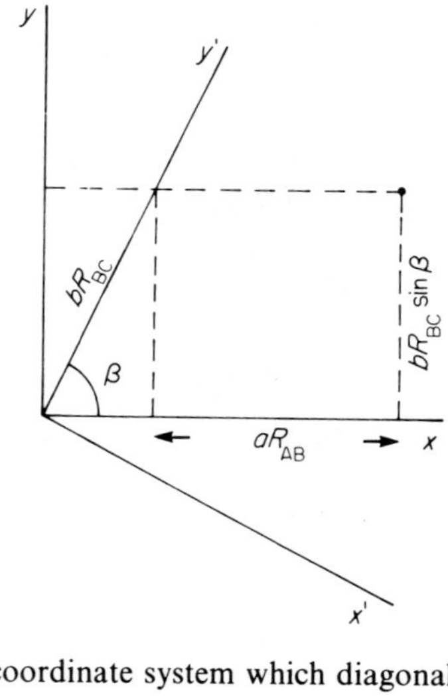

Figure 5.2 A skewed coordinate system which diagonalizes the kinetic energy

and if the parameters a, b and β are given by the expressions

$$a = \left[\frac{m_A(m_B + m_C)}{M}\right]^{1/2}$$

$$b = \left[\frac{m_C(m_B + m_A)}{M}\right]^{1/2} \qquad 5.7$$

$$\cos\beta = \left[\frac{m_A m_C}{(m_B + m_C)(m_B + m_A)}\right]^{1/2}$$

then the kinetic energy is separable as

$$T = \tfrac{1}{2}(\dot{X}^2 + \dot{Y}^2) \qquad 5.8$$

The dynamics of the reaction is therefore equivalent to that of a particle of unit mass moving under the influence of a potential $V(X, Y)$; we can think of a ball of unit mass rolling over the potential surface. In this coordinate system the reactant valley is parallel to the X axis and the product valley is parallel to the Y' axis shown in Figure 5.2. The internal coordinate in the reactant channel R_{BC} is parallel to Y, but that in the product channel R_{AB} is parallel to X'.

Figure 5.3 shows the HOCl potential plotted in skew coordinates. For this system $m_A = 35$, $m_B = 1$, $m_C = 16$, $\beta = 23.8°$. If either of the ratios m_A/m_B or m_C/m_B is small then $\beta \simeq 90°$ and the kinetic coupling term in (5.3) is not very important. This is the case, for example, when the transferred atom B is heavy compared with the entering and leaving atoms. However, if the exchanged atom is light, a so-called heavy–light–heavy system, then the skew angle is very small and there will be a very sharp bend in the reaction path leading from reactants to products, as illustrated in Figure 5.3. This will cause a trajectory coming in along the reactant valley to rise up the shoulder of the potential surface on the reactant side (the bobsleigh effect) and the product diatomic molecule will emerge with a large amount of vibrational energy. We will make a more detailed survey of the classical dynamics of such surfaces in the next section.

The skew coordinates used in Figure 5.3 do not yet provide the separation into internal and scattering coordinates which is necessary to solve the scattering problem by quantum mechanics. One approach to this which has proved successful is through the so-called natural coordinate system proposed by Marcus.† In this there is a path, called the reaction path, which starts along the bottom of the reactant valley and bends smoothly until it emerges along the bottom of the product valley. In its evolution it should pass near, but not necessarily through, the saddle point. The internal coordinates are defined to be orthogonal to the reaction path at each point on the path.

For a collinear triatomic system there is only one internal coordinate, which we call x. The value of x for a point P in this system is the distance from P to the reaction path along a perpendicular to the path, taken positive on the convex

†R. Marcus, *J. Chem. Phys.*, **45**, 4493 (1966); **49**, 2610 (1967).

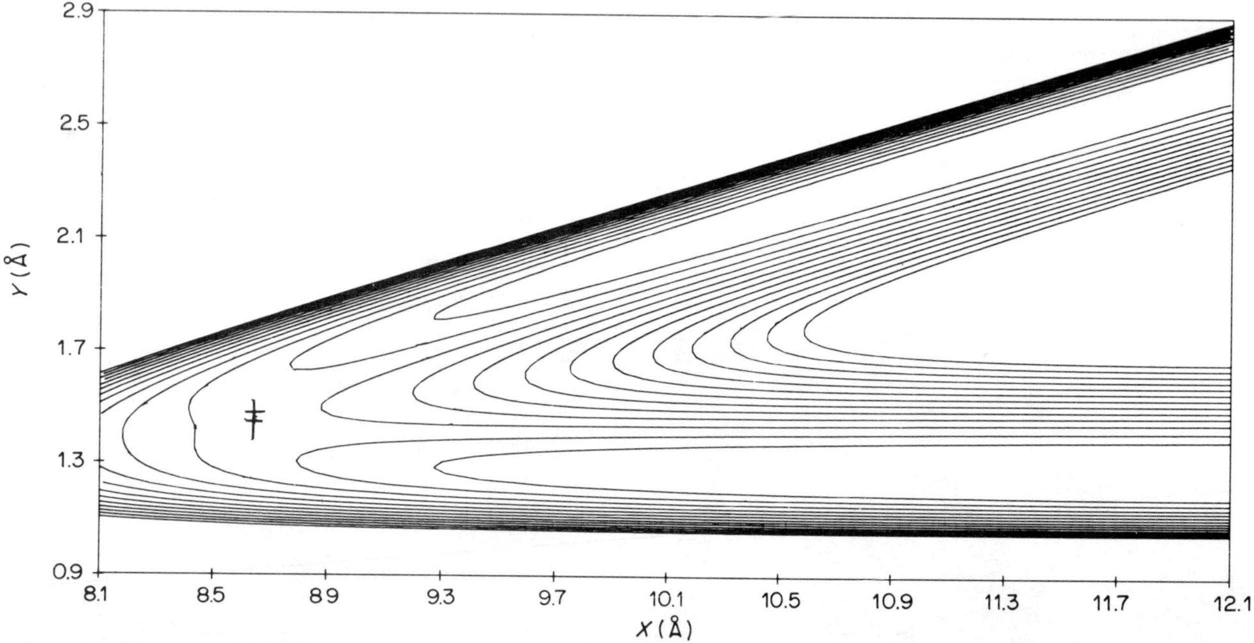

Figure 5.3 The potential of Figure 5.1 drawn in skew coordinates to remove the cross terms in the kinetic energy (equations (5.6) and (5.7)). (This figure and Figure 5.1 were kindly provided by E. Teixeira and H. Schor)

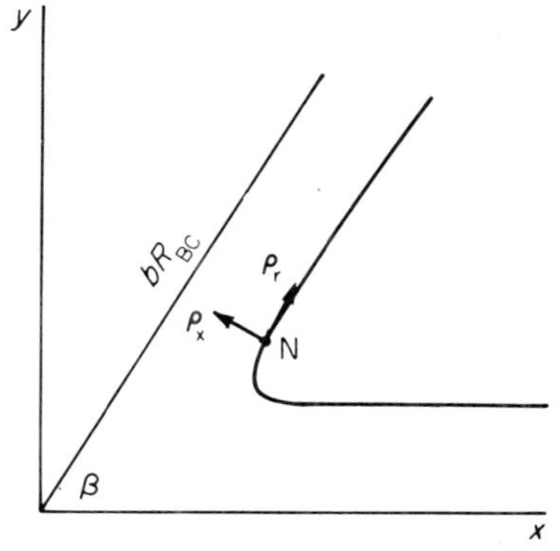

Figure 5.4 The Marcus natural reaction coordinates

and negative on the concave sides. The r coordinate of P is the distance of the perpendicular to the curve (the point N) from an arbitrary origin on the curve. An important parameter of the reaction path is its curvature $\kappa(r)$ defined by

$$\frac{\mathrm{d}\boldsymbol{\rho}_x}{\mathrm{d}r} = \kappa(r)\boldsymbol{\rho}_r \qquad 5.9$$

where $\boldsymbol{\rho}_x$ and $\boldsymbol{\rho}_r$, which are illustrated in Figure 5.4, are the orthogonal unit vectors, normal and tangent to the curve respectively, which rotate as the point N moves from reactants to products.

A disadvantage of this system is that there is an ambiguity in the coordinate of points on the concave side of the path when x is greater than the radius of curvature. A point P in this region may have two different values of x and r because it lies on two perpendiculars from the reaction path. In practice if the curvature is not large at any point the regions of ambiguity are far away from the path and of little physical significance for the dynamics.

The kinetic energy in curvilinear coordinates for a particle of unit mass can be shown to be†

$$T = \frac{\dot{r}^2}{\eta^2} + \dot{x}^2 \qquad 5.10$$

where

$$\eta = 1 + \kappa(r)x \qquad 5.11$$

In the asymptotic regions $\kappa(r)$ approaches zero and the kinetic energy is separable. In the reactant region the variables are related to those defined in the skew coordinate system (5.6) by

$$(\dot{r}, \dot{x}) = (-\dot{X}, -\dot{Y}) \qquad 5.12$$

†The analysis requires the use of differential geometry and can be found in the original papers of Marcus.

and in the product region by

$$(\dot{r}, \dot{x}) = (\dot{Y}', -\dot{X}') \qquad 5.13$$

where (X', Y') and (X, Y) are the mass weighted orthogonal coordinates shown in Figure 5.2.

By using curvilinear rather than rectilinear coordinates we have lost the separability of the kinetic energy (5.10) because the term in $\dot{r}$ depends, through η, on x. However, we have gained, not only by having a separation into internal and scattering coordinates but also because the potential energy will generally be more simply expressed in (r, x) than in (X, Y) or (X', Y'). For example, if the reaction path follows exactly the bottom of the reactant and product valleys, and passes through the saddle point, then the potential in the x variable will, to a first approximation, be a quadratic function. The curvature of this function will generally change along the path and we can encompass this through the expression

$$V(r, x) = F(r) + \tfrac{1}{2}k(r)x^2 \qquad 5.14$$

where $k(r)$ is the force constant for the vibration orthogonal to the reaction path; this passes smoothly from the BC force constant to the AB force constant as the point N moves from reactants to products. If $k(r)$ is almost constant, which may be the case in some isotope exchange reactions,

$$\mathrm{A} + \mathrm{BA}' \rightarrow \mathrm{AB} + \mathrm{A}' \qquad 5.15$$

then we might expect a vibrational state of BA′ to be conserved on exchange, i.e. to go mainly into the same vibrational state of AB. In any case expression (5.14) can be used to define an adiabatic basis, the vibrational wave functions appropriate to the potential $\frac{1}{2}k(r)x^2$, for use in a coupled-channel calculation of the reaction probabilities.

The Marcus system of natural coordinates has the advantage that it can be tailored to the potential so that full use can be made of separable approximations such as (5.14). We have already noted the disadvantages: non-separability and non-uniqueness in some regions. Other curvilinear systems have been proposed based on transformations of rectilinear variables that are independent of the potential function. One of the most important of these is that of Delves[†] which was developed for use in nuclear physics. Starting from a system such as (5.6) in which the kinetic energy is separable [Delves actually uses scaled coordinates so that the effective mass of the particle is $(m_A m_B m_C/M)$] a transformation is made to polar coordinates such that

$$X = r\cos\alpha, \qquad Y = r\sin\alpha \qquad 5.16$$

with α being defined in the range 0 to β (the skew angle). The classical kinetic energy in these coordinates is

$$T = \tfrac{1}{2}(\dot{r}^2 + r^2\dot{\alpha}^2) \qquad 5.17$$

[†] L. M. Delves, *Nucl. Phys.*, **9**, 391 (1959); **20**, 275 (1960).

Delves' coordinates in polar form are particularly suitable for describing the heavy–light–heavy triatomics for which the skew angle β is small. They can also be generalized for higher dimensions and become what are known as hyperspherical coordinates. If there are n dimensions a coordinate set can always be found in which there is only one distance variable, r, and the other $n-1$ variables are all angles. There are several slightly different definitions of these that have been explored.†

The advantage of hyperspherical coordinates is that all the reaction and product channels are at $r=\infty$ and are distinguished by different values of the angular variables. Thus r can be taken as a scattering coordinate for all reactive or non-reactive processes. The disadvantage of hyperspherical coordinates is that the contours of the potential do not follow the lines of constant r or constant angle, so there are very strong coupling terms in the potential energy for such a system.

We must, however, stress in concluding this section that the choice of variables is a problem for quantum mechanics because the wave functions have to satisfy boundary conditions which are different for different variables. In classical mechanics the matter is much less important because the equations determining the classical trajectories can be integrated numerically in any coordinate system, although there are advantages in having variables that change slowly with time. The classical treatment of reactive scattering is much the simpler and we therefore turn now to this technique and describe some of the results that have been obtained before returning to quantum mechanics in Section 5.4.

5.2 Classical trajectories for reactions

We remind the reader of the points made in Section 1.1 regarding the justification for using classical mechanics for problems in atomic and molecular scattering. We are considering particles which on an atomic scale are fairly heavy so that their quantum mechanical wavelengths are short compared with the range of their interaction potentials. In consequence the interference features which are associated with wave character may not be resolved in the scattering experiments either because of insufficient resolution in the detector or because the reactant beam has a spread of internal states and velocities.

The most extensive data on chemical reactions does not of course come from crossed beam experiments but from the traditional study of reactions in gases, liquids and solids. The liquid state has features that are outside the scope of this book, such as the diffusion rate of reactants and many-body collisions. The solid state has these and others. In the gaseous phase, however, the outcome of single collisions are interpretable within the same framework that is used for the crossed beam experiment, the only difference being that there is usually a very broad spectrum of internal states and collision velocities in the system. Gas kinetic rate constants therefore contain an average over the initial states and velocities of the

†See, for example, R. T. Pack, *Chem. Phys. Lett.*, **108**, 333 (1984).

collision partners which leaves very little scope for the quantum features to remain. Classical mechanics is therefore at its most relevant for the interpretation of such experiments. One qualification to be made to this statement is that at the threshold of reaction, when few collisions lead to reaction, quantum mechanical tunnelling through a potential energy barrier may have a significant effect on the rate constant; this is a feature we must look at later.

In Section 3.3 we looked briefly at the numerical solution of Hamilton's equations for inelastic collisions and discussed the use of action-angle variables when one or both of the colliding species is a vibrating molecule. However, as action-angle variables are internal variables one would have to choose a different set for reactants and products in a reactive scattering system and it is generally not convenient to integrate the equations of motion in a coordinate system which changes, perhaps several times, during a trajectory. Moreover, action-angle variables are not readily defined for other than diatomic molecules; in other cases more general but less elegant methods must be used.

The only other general point to be made about the integration is that, to use mathematical jargon, a mild degree of stiffness appears to be a characteristic for the equations met in classical scattering theory. The term stiff is used when stability rather than accuracy dictates the step length (time interval) of the integration method. Stability is concerned with the propagation of errors with successive integration steps. A stiff set of equations would be one for which the distance between two trajectories which differ initially by a small amount increases exponentially with the number of steps, in other words small errors propagate rapidly. Back integration of a trajectory, which can be achieved by reversing the sign of t, is a good indication of the reliability of an integrating algorithm, for if one does not return to the original conditions of the trajectory at $t = 0$, then the integration routine should be improved.

The classical dynamics of a system of N atoms is characterized by trajectories in the space of $3N$ position and $3N$ conjugate momentum variables. As the momentum of the centre of mass is a constant of the motion (in the absence of external fields) we can always work with coordinates which are referred to the centre of mass as origin so that the number of position and momentum variables will be $6N-6$. For the collision of an atom and a diatomic molecule the transformation from the eighteen space-fixed to twelve centre-of-mass fixed variables is simple, and this leads to a considerable reduction in the time needed to integrate Hamilton's equations. However, for the collision of two diatomic molecules or an atom and a triatomic molecule the transformation to centre-of-mass variables is more complicated and it is usual to omit this transformation and to work with 24 Hamilton equations rather than 18; one loses in terms of the number of equations but gains by having simpler equations. For larger systems there is even more reason to work in the space-fixed coordinate system.

Other transformations of variables could be made to reduce further the number of Hamilton equations that have to be solved. In particular the three components of the total angular momentum are constants of the motion so that, in principle, a triatomic system can be reduced to nine variables. However, the transformation is again not trivial and is consequently not used. Instead, one uses the constancy of angular momentum (also the constancy of the total energy) as a criterion that the integration is being performed accurately.

We will now examine the trajectory method for atom–diatomic molecule collisions (A + BC) in more detail, as the ability to calculate and analyse such trajectories has been a major step in our understanding of elementary chemical reactions. We concentrate here on the principles of the method and give details of the dynamical equations in Appendix 6; these are based on the treatment by Karplus, Porter and Sharma.[†]

In the classical trajectory method one examines a batch of trajectories, each with different initial conditions, and makes a statistical analysis of the outcome. This is usually called the Monte Carlo method. The number of trajectories that have to be examined depends on the questions being asked. For example, if one takes a batch of N trajectories, all conforming with equal weight to certain criteria for their initial conditions ($t = 0$), and if N_r lead to reaction and $N - N_r$ to no reaction, then the probability of reaction for these initial conditions is

$$P = \frac{N_r}{N} \qquad 5.18$$

and the standard error of this result is

$$\varepsilon = \left(\frac{N - N_r}{NN_r}\right)^{1/2} \qquad 5.19$$

which we note is roughly proportional to $N^{-1/2}$ if N_r is a reasonably large fraction of N. To double the accuracy of a result one needs to increase the number of trajectories by a factor of four.

If N_r is a small fraction of N the error is proportional to $N_r^{-1/2}$; hence one needs more trajectories to obtain a given level of accuracy for improbable events. For example, if one is interested in knowing what is the probability of a product being in a particular vibrational and rotational energy state, then $N_r(v, j)$ is generally a small fraction of N and the error in

$$P(v, j) = \frac{N_r(v, j)}{N} \qquad 5.20$$

will be large.

Let us now examine the initial conditions for an A + BC trajectory. In the centre-of-mass coordinate system we have twelve conditions to be specified. Some of these will be determined by the initial state of the reactants, some will be arbitrary (arising from an arbitrary choice of axes) and others will be chosen randomly so as to simulate the whole range of collisions that can occur.

In an ideal state-selected crossed beam experiment BC would be in a well-defined vibrational–rotational state so the quantum numbers v and j will

[†]M. Karplus, R. N. Porter and R. D. Sharma, *J. Chem. Phys.*, **43**, 3259 (1965).

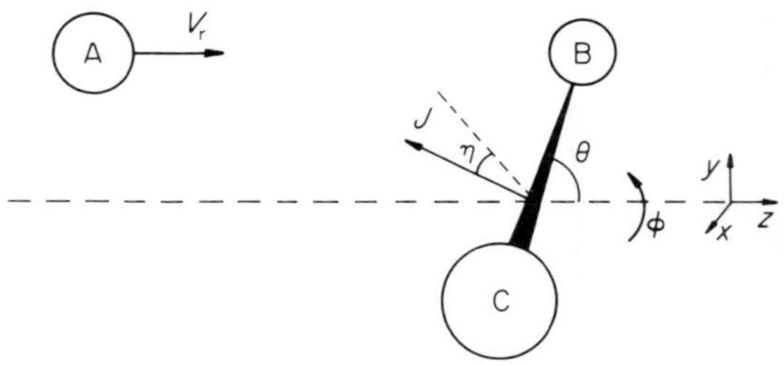

Figure 5.5 Parameters for atom–diatomic molecule trajectories

determine two of our twelve initial conditions. A third is given by the relative velocity of A to the centre of mass of BC. This is obtained from the cosine rule

$$v_r = (v_A^2 + v_{BC}^2 - 2v_A v_{BC} \cos\gamma)^{1/2} \tag{5.21}$$

where v_A and v_{BC} are the beam velocities and γ is the beam crossing angle.

There is no loss of generality in placing A and the centre of mass of BC in the *yz* plane with v_r parallel to the *z* axis, as shown in Figure 5.5. We can, therefore, arbitrarily put two initial momenta as zero (the components of μv_r in the *x* and *y* directions) and one position coordinate (*x*) is also initially zero.

The initial value of the *y* relative position coordinate is called the impact parameter, *b*, by analogy with the central force problem shown in Figure 2.1. In a real collision any value of *b* is possible but the probability of reaction will fall to zero as *b* goes to infinity. In general there is a sharp decrease in the probability of reaction beyond a certain value of *b* and by running a small number of trajectories one can establish a *minimum* value of *b* beyond which no reaction occurs; this is typically a few angstroms. This minimum value gives the upper limit of *b*, called b_{max}, which has to be sampled in a batch of trajectories in order to simulate all feasible reactive collisions.

The usual procedure for selecting the *b* value for a trajectory is to generate a random number, ξ, between zero and one and to take

$$b = \xi b_{max} \tag{5.22}$$

This gives a batch of trajectories with a uniform random sample of *b* values between 0 and b_{max}. In a physical situation an impact parameter *b* occurs with a weighting proportional to *b* (the area of the cross-section between *b* and $b + db$ is $2\pi b\,db$) so that with this sample one must weight the result of each trajectory by its *b* value. An alternative procedure is to construct a sample from

$$b = \xi^{1/2} b_{max} \tag{5.23}$$

and then give each trajectory equal weight. However, this procedure gives more trajectories with high *b* values, and as these have the lowest probability of reaction, it is generally less efficient than using (5.22).

Although the magnitude of the rotational angular momentum of BC is defined by *j*, its direction is not specified and can take any angle (η) between 0 and 2π with respect to a reference normal to the BC axis (the dashed line in Figure 5.5). The initial orientation of BC with respect to the space-fixed axes (Figure 5.5), which is defined by two polar angles (θ, ϕ), is also not determined. We therefore have three angles associated with the rotation of BC whose initial values have to be determined by random number generation.†

The reader might wish to check that the number of variables whose initial values have been considered in the above discussion is ten; hence there are two still to be considered. These are the initial phase of the vibrating molecule and the initial distance in the z direction between A and BC. The latter is clearly unimportant, providing it is sufficiently large that the interaction potential between A and BC can be neglected. We could therefore take an arbitrary value for this distance, d say, and choose the vibrational phase randomly. In practice it is easier to take a fixed phase for the oscillator and to chose a random value (outside the range of the potential) for the A–BC distance. This procedure is entirely equivalent to randomizing the oscillator phase.

If the time taken for BC to go through a complete vibrational cycle is τ (roughly independent of v and j), then in this time the A–BC z value will be reduced by τv_r. It follows that we can select the initial z value to be in the range $-d$ to $-(d + \xi\tau v_r)$, where ξ is another random number, and have the oscillator in a random phase by the time the z value equals $-d$. The reason that this procedure is simpler is that we can choose the oscillator phase to be 0 or π so that the B–C distance is either at its minimum (R_-) or its maximum (R_+) value. There is then no momentum along the BC bond and hence no cross term (Coriolis force) between vibration and rotation; all the BC kinetic energy is in rotation. The only complication with this procedure is that one generally needs an iterative procedure to find R_- and R_+, the classical turning points of the effective potential (see Figure 2.2) for each (v, j) state. It is through $R_\pm$ that the initial vibrational energy of BC enters the calculations.

The outcome of a trajectory may be that no reaction has occurred or that products B + AC or C + AB have formed, or at higher energies one might obtain three separate atoms. It is therefore necessary to calculate the values of the three internuclear distances during the course of the trajectory and to terminate the calculation when at least two of these exceed some specified values. If the collision occurs over a very attractive interaction potential, i.e. ABC is a stable molecule species, then it is possible that a trajectory may become trapped in the interaction region, and although this trajectory will eventually emerge, at least back to the reactant channel if not to products, the loss of numerical accuracy over a long trajectory may make the outcome not significant. The trajectory programme will therefore usually have a termination instruction based on time as well as on the interatomic distances.

When the trajectory is completed the coordinates and momenta will be

†The weighting given to each value of these angles must be considered carefully. For example, of the two polar angles determining the orientation of BC, one is uniform from 0 to 2π and the other is weighted by sine of the angle.

transformed to the appropriate products and analysed to give the distribution of energy between the relative translational motion and the vibrational and rotational motion of the diatomic species. The scattering angle in the centre-of-mass coordinate system will be given by the angle between the incoming vector $\mathbf{v}_r$ and the equivalent outgoing vector for the relative translational motion. From the vibrational and rotational energies we can determine semi-classical quantum numbers v' and j', and these can be discretized by rounding them to the nearest integer number.

The probability of reaction to particular products for specific values of v_r, v, j and b is called the opacity function, P_r. This is the ratio of the number of trajectories leading to specific products to the total number of trajectories, with all initial values other than those specified averaged, or given arbitrary values, according to the discussion given above. The reaction cross-section for this initial state is

$$\sigma_R(v_r, v, J) = 2\pi \int_0^{b_{max}} P_R(v_r, v, J, b) b \, db \qquad 5.24$$

which is just the sum of the reaction probabilities for each annulus of area $2\pi b \, db$, from $b = 0$ to $b = b_{max}$.

Expression (5.24) is the cross-section for reactions that proceed over the particular potential energy surface on which the trajectories have been calculated. There is, however, a further complication; for most reactive systems there is more than one surface that emanates from the state of the reactants and the total cross-section is the appropriate average for motion on all of these. For example, in the reaction

$$Cl(^2P) + O_2(^3\Sigma_g^-) \rightarrow ClO(^2\Pi) + O(^3P) \qquad 5.25$$

the total electronic degeneracy of the reactants is six for Cl (twofold spin multiplied by threefold orbital degeneracy) and three for O_2 (all spin), so that there are eighteen different electronic states that are asymptotic to these reactants. The ground-state surface of ClO_2 which dissociates to these reactants has a doublet spin degeneracy but no orbital degeneracy, so that only one in nine collisions of the reactants will pass over the ground-state surface. The other eight will lead to excited states of ClO_2. If the reaction (5.25) can only proceed over the ground-state surface, which would be the case if all the excited-state surfaces had high barriers to the formation of products, then to compare the calculated cross-section with experimental quantities one would have to multiply (5.24) by a degeneracy factor, which in this case is 1/9. If products could be formed over excited-state surfaces then we would have to calculate trajectories on these and multiply by the appropriate electronic degeneracy factors.

More than one electronic surface may also be involved for non-reactive inelastic collisions, but most experiments in this field refer to closed-shell molecules for which there is only a single surface asymptotic to the reactants.

Complete state selection of the reactants is very difficult in a crossed beam experiment. It is usually not difficult to lower the effective temperature of the beams so that only the ground vibrational states of the reactants are significantly populated, but excited rotational states cannot usually be neglected. To determine cross-sections for different sets of reactant states one therefore aims not to have pure state-selected beams but to change the populations in a controlled manner and by this to deduce state-selected cross-sections.

If there are reactant molecules with non-zero rotational quantum numbers J, then it is possible to achieve some selectivity in the $2J + 1$ components of these states (effectively orienting the molecules with respect to the beam axis) by using molecular beam equipment containing an electrostatic focusing chamber. This is most easily achieved for polar symmetric top molecules because these have a first-order Stark splitting, the energy components of the rotational state $|JKM\rangle$ being split by an amount proportional to the product of K, the component of $\mathbf{J}$ along the figure axis, and M, the component of $\mathbf{J}$ along the electric field, divided by $J(J + 1)$. Such molecules can be selectively focused in a hexapolar field (one whose cross-section changes sign six times around the axis) plus a weak orientating dipolar field. By noting the change in reactivity when the direction of the orienting field was reversed, one of the classic results of state-selected chemistry was obtained, which is that the reaction

$$K + CH_3I \rightarrow KI + CH_3$$

occurs when K collides with the iodine atom, and not when it collides with the CH_3 group.[†] This is not of course a surprising conclusion, but the technology required to prove it was formidable.

With continuing improvement in the experimental techniques for state selection of reactants there is an obvious incentive for performing trajectory calculations which lead to predictions of state-selected cross-sections such as (5.24), or even cross-sections for specific orientational components of the rotational states.

In any experimental situation there will be some averaging over v_r, v and j; we have already stressed that even in a crossed beam experiment there is a spread of values for v_r and usually imperfect state selection. It follows that cross-sections such as (5.24) must be weighted by state populations in order to compare calculated and experimental results. In an equilibrium gas, for example, the fraction of molecules in particular (v, j) states is given by the Boltzmann distribution

$$f(v, j) = Q^{-1} g_{vj} \exp\left(-\frac{E_{vj}}{kT}\right) \qquad 5.26$$

where E_{vj} is the energy of the state, g_{vj} its degeneracy and Q the vibration–rotation partition function. This distribution could, for simplicity, be obtained from the harmonic oscillator rigid-rotor model for BC.

The distribution of relative velocities in an equilibrium gas is given by the

[†]R. H. Beuhler, R. B. Bernstein and K. H. Kramer, *J. Am. Chem. Soc.*, **88**, 533 (1966).

Maxwell–Boltzmann distribution

$$f(v_r) = 4\pi\left(\frac{\mu}{2\pi kT}\right)^{3/2} v_r^2 \exp\left(-\frac{\mu v_r^2}{2kT}\right) \qquad 5.27$$

The rate constant for a bimolecular reaction is a macroscopic quantity which can be obtained from the calculated reaction cross-section by averaging over the internal states of BC and the relative velocity v_r. A 'microscopic' rate constant can be obtained from σ_R by multiplying by the relative velocity

$$k(v_r, v, j) = g^{-1} v_r \cdot \sigma_R(v_r, v, j) \qquad 5.28$$

where g is the electronic degeneracy. It can be seen that this expression is dimensionally correct but the detailed analysis is not trivial.† The velocity-averaged rate constant is

$$k(T) = \int dv_r \sum_{vj} k(v_r, v, j) f(v,j) f(v_r) \qquad 5.29$$

where the functions f are given by expressions (5.26) and (5.27).

5.3 Potential energy functions and reaction cross-sections

We have made very few references in this book to the mathematical representation of potential energy functions. For atom–atom collisions there are a large number of these: Morse, Lennard–Jones, etc. For inelastic atom–diatomic molecule collisions there are again a large number of possibilities and quite frequently the Legendre expansion (4.70) is used with the functions $K_\lambda(r)$ being atom–atom potential functions. For non-reactive molecule–molecule collisions the functions become more complicated; they have more dimensions and there is less agreement on their general form. One can make bipolar expansions about the centres of mass; effectively this is a generalization of (4.70). Another common assumption is that the potential is a pairwise sum of atom–atom potentials. An even more difficult problem is to devise a potential for a reactive scattering situation because the whole, or a large part, of the $(3N-6)$-dimensional space of internal coordinates may have to be represented. There are some general approaches to this problem but we must refer the reader to other sources for the details, and selected references in this field are given in the Biliography.

Even if one does not have a mathematical representation of the potential it is clear that there may be some topological features of the potential surface that determine the dominant features of a reaction cross-section or even of a chemical rate constant. In the chapter on rotational inelastic scattering we presented arguments for the importance of molecular shape, as approximated by a hard potential, in determining rotational rainbows. For reactions the potential energy functions have much richer variety, but in many cases the most important feature

†See, for example, J. C. Polanyi and J. L. Schreiber, in W. Jost (ed.), *Physical Chemistry—An Advanced Treatise*, Vol. 6A, p. 389. London, Academic Press, 1974.

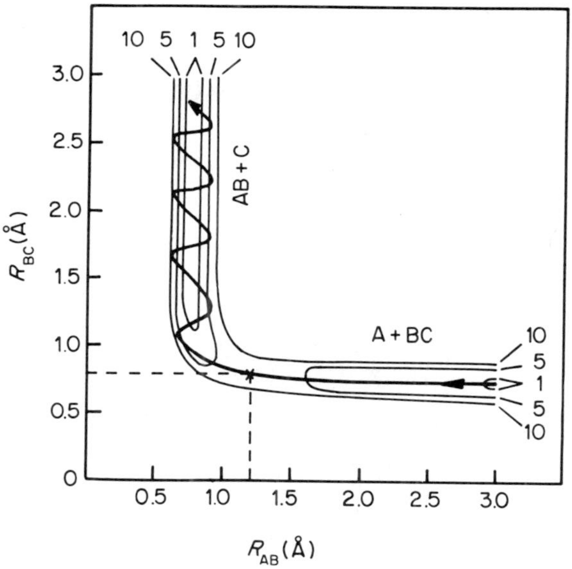

Figure 5.6 A typical trajectory for a collinear A + BC collision on a surface for which the barrier is in the reactant valley. The contours are labelled by their relative energies in kcal/mol. (*Data for this, and for Figure 5.7, are taken from J. C. Polanyi and W. H. Wang, J. Chem. Phys.*, **51**, *1439, 1969*)

is the saddle point on the reaction coordinate; this we have already mentioned in Section 5.1 in relation to the Marcus system of natural coordinates. There is a close relationship between the barrier height along a reaction path and the experimentally determined activation energy. This is central in the statistical theory of chemical reactions where the saddle point is usually identified with the transition state. The geometry of the transition state has also a very important influence on the rotational and vibrational energy distribution in the products. Although such relationships had been postulated in early work in chemical kinetics, it was not until classical trajectory calculations became available in the 1960s that detailed analyses were possible, the most extensive work in this field being by J. C. Polanyi and his coworkers.

Figure 5.6 shows a typical trajectory for a collinear A + BC thermally neutral reaction on a potential surface where the barrier along the reaction path occurs in the reaction valley; i.e. the highest point on the lowest energy path leading from reactants to products occurs when A has approached B but before the BC bond has been lengthened.*

The barrier height on this surface is 0.3 eV and the initial translational energy of 0.39 eV is just sufficient to surmount this barrier; there is no initial vibrational energy in the reactant BC molecule. For this trajectory 79 per cent of the

*The trajectory has been calculated for atoms of equal mass so that if the potentials had been plotted in skew coordinates (to remove off-diagonal kinetic energy terms), the skew angle would be 60°.

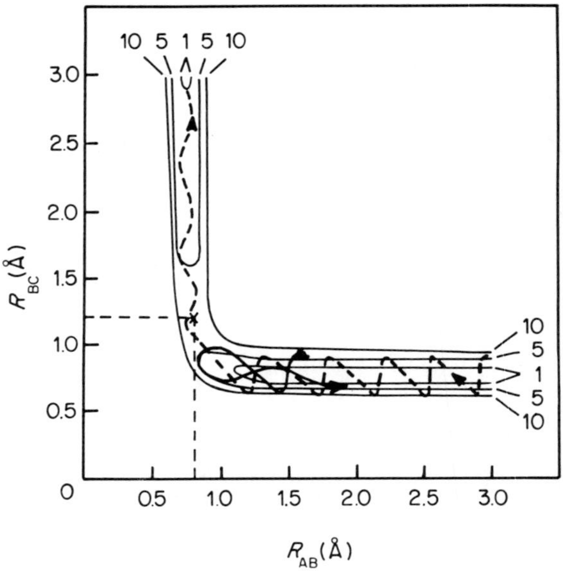

Figure 5.7. Typical trajectories for a collinear collision on a surface for which the barrier is in the product valley

translational energy of the reactant translational energy is converted into vibrational energy of the product AB, leaving just 21 per cent for the translational energy of AB relative to C. If the reactant translational energy is increased to 0.69 eV, then it is found that almost 100 per cent of this is converted to vibrational energy of the products. In contrast, Polanyi and Wang showed that transferring some of this translational energy into vibrational energy of BC inhibits the reaction.

The potential energy surface shown in Figure 5.6 is quite similar to that for the $F + H_2$ reaction, except that this is exothermic. We can therefore understand why this reaction produces a lot of vibrational energy in the product HF; in fact this system is a chemical laser as there is an inverted (non-Boltzmann) population of these vibrational levels.

As trajectories are reversible it is clear from Figure 5.6 that if we were to start from the product valley with the same total energy then we must put most of the initial energy into vibrational rather than translational motion in order to surmount the barrier. Whether or not the trajectory passes over the barrier will depend on the phase of the vibration of the diatomic molecule. Figure 5.7 shows a surface in which the saddle point is in the exit valley having two trajectories with the same vibrational (0.33 eV) and translational (0.065 eV) energies but different initial vibrational phases. One of these leads to reaction, the other does not.

Although Figures 5.6 and 5.7 have not been drawn with skew axes to take account of the mass ratios of the atoms, it is clear that these ratios will also have a strong influence on the reaction cross-sections. Broadly speaking, if the skew angle, given by expression (5.7), is sharp e.g. in a heavy–light–heavy combination of atoms, then the trajectories may spend a lot of time in the bend of the potential, as the light atom bounces back and forth between the two heavy atoms, and the chance of reflection rather than reaction is increased. However, in such a situation the collinear model of the reaction may give results that are quite different from those of three-dimensional collisions in which the initial impact of A and B is likely to lead to rotation of the BC molecule.

The potential on which the trajectories shown in Figures 5.6 and 5.7 are based has a collinear transition state, and the barrier to reaction increases rapidly as the angle $A\hat{B}C$ decreases. For this reason we expect little rotational energy in the product BC unless the collision energy is considerably greater than the barrier height, and indeed at a collision translational energy of 0.39 eV three-dimensional trajectories produce only 2.3 per cent of the energy release as rotation. Even when the collision energy goes up to 0.69 eV, only 15 per cent of the product energy is in rotation.

The relationship between the potential energy function and rotational energy in the reactants or products is not, however, a simple matter. The total angular momentum is a constant of the motion so that if **L** is the orbital angular momentum of the reactants in an A + BC collision and **j** is the angular momentum of the diatomic molecule, then the total angular momentum is

$$\mathbf{J} = \mathbf{L} + \mathbf{j} \qquad 5.30$$

which by the triangle rule gives the following bounds on the magnitude of **J**:

$$|L - j| \leqslant J \leqslant L + j \qquad 5.31$$

There will be similar bounds for the angular momenta of the products AB + C:

$$|L' - j'| \leqslant J \leqslant L' + j' \qquad 5.32$$

If the saddle point has a similar geometry to that shown in Figure 5.8, and the entrance and exit valleys are rather narrow, then the reactive trajectories may all pass along a valley in which the angle α is nearly constant. In this case L and L'

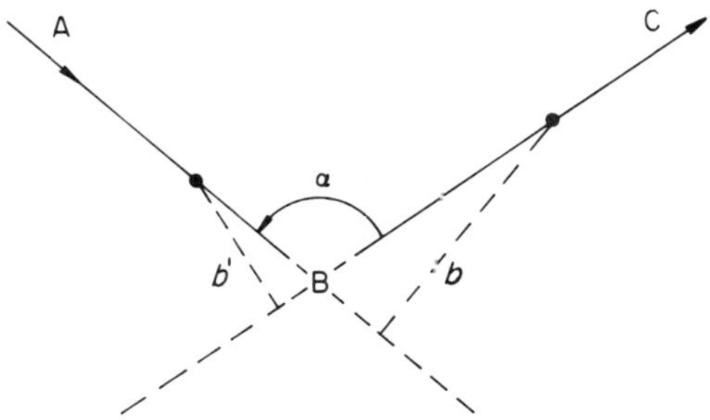

Figure 5.8 Relationship between impact parameters in the entrance and exit valleys and the geometry of the saddle point. The dots show the centres of mass of AB and BC

have a narrow range, being determined by the impact parameters b and b', both of which are proportional to $\sin\alpha$. The distribution of rotational states of the reactant BC is determined by the Boltzmann probability factor for an equilibrium gas, and for a light diatomic molecule, say H_2 or HX, the maximum value of j will not be large. This means that J may have a rather narrow range, typically from $L - j_{max}$ to $L + j_{max}$. The maximum value of j', the angular momentum of the product AB, will then be $L + L' + j_{max}$ and the minimum value will be approximately zero, if L and L' are not too different.

If the entrance and exit valleys are broad or the collision energy is high, then there will be a larger range of impact parameters leading to reaction, and the bounds (5.31) and (5.32) give little help in deducing the role of rotation energy in the reactants or its distribution in products.

For a general A + BC scattering experiment with unoriented reactants the product distribution is fully defined by three angular variables. These are the scattering angle θ between the initial $\mathbf{k}$ and final $\mathbf{k}'$ momentum vectors, which we have already discussed, the orientation angle χ between $\mathbf{k}$ and $\mathbf{j}'$, and the dihedral angle ϕ between the $\mathbf{k}$, $\mathbf{k}'$ and $\mathbf{k}, \mathbf{j}'$ planes. The triple-angle correlation function among $\mathbf{k}, \mathbf{k}', \mathbf{j}'$ is described by a probability function. $W(\theta, \chi, \phi)$, and as Herschbach has pointed out,† this contains maximum information about the directional characteristics of the experiment. The probabilities of finding angles in an incremental range are

$$W(\theta, \chi, \phi) \sin\theta \sin\chi \, d\theta \, d\chi \, d\phi \qquad 5.33$$

Although the product vectors $\mathbf{k}'$ and $\mathbf{j}'$ have azimuthal symmetry about $\mathbf{k}$ it is still possible to determine the ϕ distribution by measuring $\mathbf{k}'$ and $\mathbf{j}'$ simultaneously, i.e. determining the correlation in their azimuthal distributions.

In a crossed beam experiment one determines the angular distribution of products. If a time-of-flight detection method is used their velocities, and hence their translational energies, can also be determined. It is not a trivial matter to convert this into a centre-of-mass coordinate system for comparison with theoretical predictions. Although the essentials of this transformation have been given in Section 1.5, there is the practical difficulty that the experiment never has ideal monoenergetic beams. There are frequently substantial statistical errors because the experimentalist is usually aiming for the maximum angular resolution and this implies having the minimum significant count rate in the detector.

The transformation from laboratory to centre-of-mass coordinates can be expressed mathematically by the integral

$$I_{lab}(v_1', \theta, \phi) = A \int dv_1 \int dv_2 n_1(v_1) n_2(v_2) v_r \mathfrak{J}(v_1, v_2/v_r, V) I_{cm}(v_r', \theta) \qquad 5.34$$

where I_{lab} is the measured laboratory angular distribution of a detected product of the reaction with velocity v_1' and I_{cm} the centre-of-mass angular distribution

†D. R. Herschbach, *Disc. Faraday Soc.*, **38**, 283 (1962).

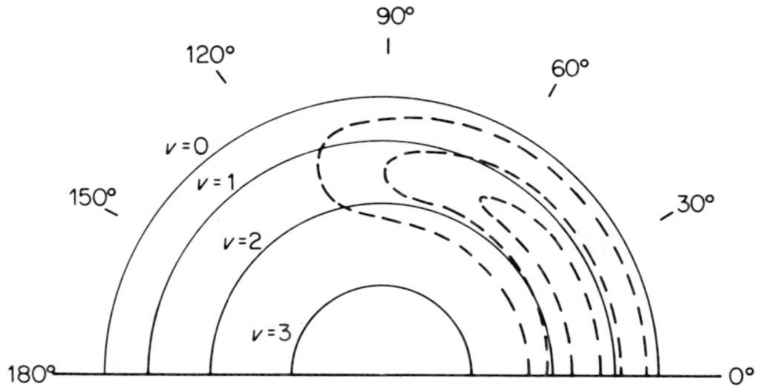

Figure 5.9 Idealized centre-of-mass flux–velocity contour map

with velocity v_r'. The relative velocity of the incident beams is v_r, and $\mathfrak{J}$ is the jacobian transforming the centre of mass to laboratory coordinates. The velocities of the incident beams are v_1 and v_2, and $n(v)$ their number distributions. A is a normalizing constant. The mathematical technique available to extract I_{cm} from a measured I_{lab}, a so-called deconvolution, has been described by Siska,† and we do no more than emphasize that it is not a trivial matter.

$I_{cm}(v_r', \theta)$ can be plotted in what is called a centre-of-mass flux–velocity contour map, treating v_r' as the radial variable. An idealized map is shown in Figure 5.9. The contours at greatest distance from the origin are those for products having the largest velocity, and hence the lowest internal energy. For example, if in a reaction

$$A + B_2 \rightarrow AB(v, j) + B$$

there was isotropic scattering then the contours for each (v, j) state would appear as a circle on such a map with the (0, 0) state having the largest radius.

There are three important limiting patterns of the centre-of-mass differential cross-section: forward scattering, backward scattering and symmetric scattering. The latter, a forward–backward symmetry about $\theta = 90°$, is evidence that the reaction goes through a complex which has a lifetime which is long compared to its rotational period. The complex therefore undergoes several revolutions before dissociating. However, even if there is forward–backward symmetry the scattering cross-section is not isotropic, but depends on the moments of inertia of the complex. For example, if it is a symmetric top ($I_b = I_c$) and prolate (cigar-shaped) ($I_A < I_B = I_C$), then the cross-section peaks at $\theta = 0$ and 180°. If it is an oblate (discus-shaped) top ($I_A > I_B = I_C$), then it peaks at $\theta = 90°$.††

If there is pronounced asymmetry in the forward–backward scattering pattern, then we can rule out the formation of complexes as a significant part of the reaction. Surfaces of the type illustrated in Figure 5.6 will give this behaviour.

†P. Siska, *J. Chem. Phys.*, **59**, 6052 (1973).
††W. B. Miller, S. A. Safran and D. R. Herschbach, *Faraday Disc.*, **44**, 108 (1967).

However, if a potential surface has a minimum, which may be stable or metastable to dissociation, it does not necessarily mean that the scattering pattern will show evidence of complex formation because the reactive trajectories may not all pass through that region of the potential. Also, at high collision energies most trajectories are likely to pass rapidly through any minima and complex formation is again unlikely.

The classical examples of backward scattering with the reaction cross-sections being predominantly in the range 90–180° are of alkali metals with methyl iodide. Reaction in these systems arises mainly from trajectories with low impact parameters so that the products rebound and the reactions usually have small total cross-sections.

In forward scattering (0–90°) reaction occurs with trajectories having large impact parameters, and the total cross-section will therefore be large. The reactions of alkali metals with halogens fall into this pattern and are called stripping reactions. Such reactions are usually exothermic and have no barriers. In the reaction of K with Br_2, for example, the potential surface is flat until a K–Br distance is reached at which electron transfer occurs, to give $K^+Br_2^-$, and after

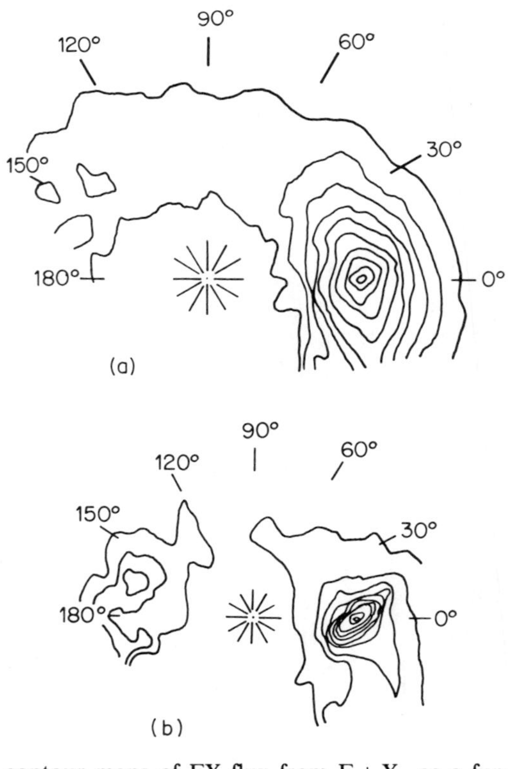

Figure 5.10 Polar contour maps of FX flux from $F + X_2$ as a function of the centre-of-mass scattering angle.The incident F atom is at $\theta = 0°$ and the incident X_2 at 180°. (a) X = Br, (b) X = Cl. (*After N. C. Firth and R. Grice*, Mol. Phys., **60**, *1261, 1273, 1987*)

this K^+Br^- forms exothermically with ejection of a Br atom. This has been called the harpoon mechanism.

Figure 5.10 shows flux–velocity maps for two halogen atom–halogen molecule reactions studied by Firth and Grice. One of these figures, that for $F + Cl_2$, shows sizable amounts of both backward and forward scattering and the other, $F + Br_2$, is almost entirely forward scattering. The former has been interpreted as due to a complex FCl_2, and from the scattering pattern it is concluded that this is roughly a prolate top with a $Cl\hat{C}lF$ bond angle of approximately 150°. In contrast, at the collision energy of the experiment, there is no evidence for the formation of a complex FBr_2.

One of the important objectives of trajectory studies is to calculate rate constants, or other measured quantities of a reaction, and to say which of several different experimental measurements are likely to be the most reliable. Differences occur because there are often several ways of studying a reaction and there are usually some assumptions behind the resulting rate constants. For example, there are very few elementary gas phase reactions that are not accompanied by other reactions, and frequently a rate constant of interest depends on assumptions about the values of other rate constants in the reaction scheme.

Not surprisingly the most important requirement in this type of trajectory study is to have an accurate potential energy function, but it has to be admitted that at present there are few systems where one has total confidence in the potential. Although the classical trajectory mehod has its own limitations, one can usually be certain that failure to agree with a well-established experimental quantity is due to an error in the potential energy function rather than to limitations of the classical trajectory method. We illustrate this type of calculation by a practical example from our own studies.

The reaction

$$Cl(^2P_{3/2}) + O_3(^1A') \rightarrow ClO(^2\Pi) + O_2(^3\Sigma_g^-) \qquad 5.35$$

is part of the catalytic cycle for chlorine atom depletion of ozone and oxygen atoms, which is important for upper atmosphere chemistry. A potential function has been derived for the ground state (^2A) of ClO_3 by the many-body expansion method.† In this the potential, relative to the energies of the separate atoms, is taken as a sum of the interaction energies of the atoms in pairs, plus the interaction energies of the atoms in threes, plus a four-body term. From the two-body and three-body terms alone it was found that the reaction proceeded by the approach of Cl to O_3 along a path collinear to one O–O bond with an activation energy of about 0.31 eV. This is considerably greater than the experimental measurements of the activation energy and the four-body term was therefore chosen to reduce this barrier to about 0.015 eV, which appears more consistent with experimental estimates, and to reproduce an estimated energy for the known

†This is described fully in J. N. Murrell, S. Carter, S. C. Farantos, P. Huxley and A. J. C. Varandas, *Molecular Potential Energy Functions*, John Wiley and Sons, 1984.

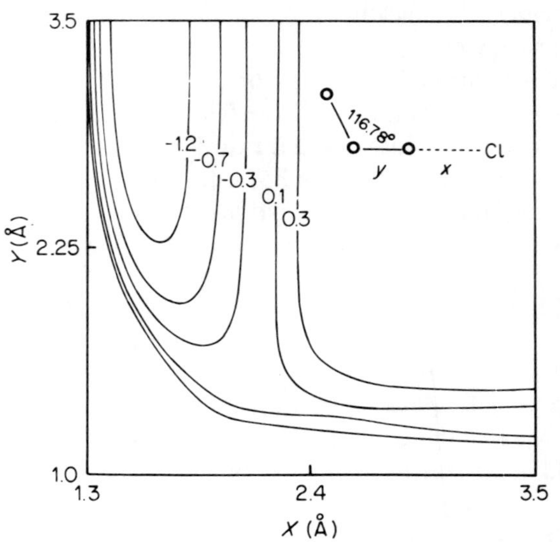

Figure 5.11 Contours of the ClO_3 potential for the planar geometry shown. The length of the leaving O_2 has been optimized at each value of x and y. (*Data from S. C. Farantos and J. N. Murrell,* Int. J. Quant. Chem., **14**, *659, 1978*)

stable species ClO_3. Figure 5.11 shows contours of the ClO_3 potential covering the reaction path.

Trajectories were calculated at four different collision energies corresponding to the mean energies of a maxwellian distribution of velocities between 200 and 600 K with the O_3 in its ground vibrational state. The initial rotational energy of O_3 was given the equipartition value $3kT/2$ and its angular momentum was chosen randomly consistent with this energy. A degeneracy factor of $\frac{1}{2}$ was adopted because the reactants in (5.35) give rise to two doublet surfaces, and it is probable that only one of these leads to products at the collision energies of interest.

Figure 5.12 shows the calculated cross-sections $\sigma(E_r)$ at four collision energies and a linear extrapolation to a threshold energy. From these values rate constants can be calculated using expressions such as (5.28) and (5.29). At room temperature the result is 1.34×10^{-11} cm^3 molecule^{-1} s^{-1}, which compares well with the experimental value of 1.2×10^{-11} cm^{-3} molecule^{-1} s^{-1}. However, trajectory calculations of this kind are very sensitive to the barrier height on the reaction path, which in this case was chosen to give a rate constant agreeing with experiment. We should therefore say that the calculations are consistent with the experiments rather than giving conformation of their accuracy.

Other results from the trajectory calculations have not yet been tested by experiment. It is predicted that at 300 K the energy release in the reaction, which is nearly 2 eV, should be 57 per cent in ClO (which is almost equally shared between vibration, rotation and translation) and 43 per cent in O_2 (4 per cent vibration, 9 per cent rotation and 30 per cent translation). The large amount of

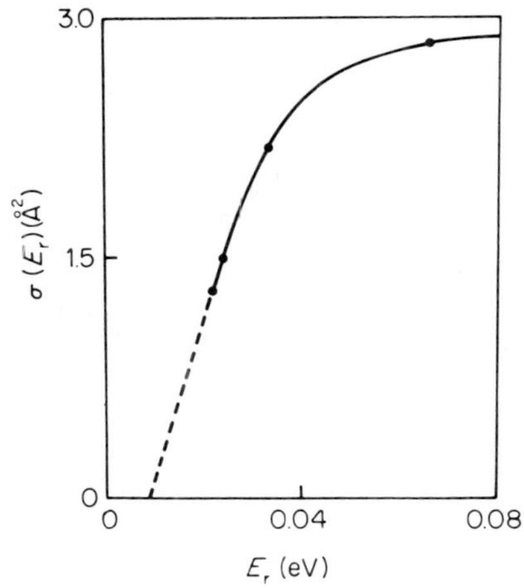

Figure 5.12 Cross-section for reaction (5.35) as a function of the collision energy

vibrational energy in ClO can be understood from the position of the transition state in the entrance valley (Figure 5.11). There is much less vibrational energy in O_2 because the bond length does not change much in the reaction.

There has not yet been a crossed molecular beam study of the reaction so there is no measurement of the differential cross-section. At 300 K it is calculated that in the centre-of-mass system the ClO is produced predominantly in the backward-scattering direction and the O_2 in the forward direction. The large forward–backward asymmetry shows that complex formation is not important, and an examination of trajectories shows that the stable species ClO_3, which is a minimum on the potential surface, are not formed in the reaction. Insufficient trajectories were run to produce statistically significant flux–velocity diagrams.

5.4 Quantum theory of reactive scattering

We have already stressed in the introduction to this chapter that the main problem we face in the quantum theory of reactive scattering is that in different reaction channels we require different coordinates and different basis sets for the internal states. Although we can formally write down multichannel equations in a generalized reaction coordinate, using hyperspherical coordinates, for example, to calculate scattering matrices we must always transform the scattering solutions in the asymptotic regions to internal basis functions appropriate to the particular reaction channel.

Another problem we face in reactive scattering is that there are frequently a large number of open channels. A high collision energy is often needed to surmount a potential energy barrier on the reaction path, and this may lead to a large energy release in the reaction channels with the product molecules produced in a wide range of rotational and vibrational states. This will be particularly prominent for exothermic reactions. In consequence we need a large number of channels, open and perhaps closed, to obtain covergence in the scattering matrices.

It is therefore perhaps not surprising that there have been few reactive systems that have been treated quantum mechanically. The notable exception to this statement is the $H + H_2$ reaction. This at least has the merit that the rotational energy spacing of the reactants and products is high, so that at collision energies just sufficient to surmount the potential energy barrier there will be relatively few open channels. $H + H_2$ and its isotopic variants have become, in the collinear, coplanar (two-dimensional) and full three-dimensional modes, benchmarks for reactive scattering calculations.

We turn first to an examination of collinear A + BC reactive collisions. This model was first introduced in Section 3.5 for vibrationally inelastic collisions, and it has the merit of being mathematically simple yet retaining the basic physics of the problem. However, it must be stressed that the methods used for its solution cannot always be generalized to two or three dimensions.

The first accurate reaction probabilities for collinear $H + H_2$ using the coupled-channel method were obtained by Diestler.† He used as a reaction coordinate the distance between the two most distant atoms (r_{AC}) and as internal coordinates r_{BC} in the reactant and r_{AB} in the product valleys. The wave function can be expressed as a sum of two functions, the first expandable in the BC vibrational basis and the second in the AB vibrational basis. The resulting coupled equations can then be integrated from $r_{AC} = 0$ to $r_{AC} = \infty$, and from a set of n independent solutions one can obtain appropriate asymptotic forms of the scattering wave functions, in a similar way to that described for the vibrationally inelastic problem in Chapter 3. Note that this coordinate system is not easily generalized to three dimensions, particularly if the reactive channel AC + B is open.

Using a basis of up to five vibrational states for reactants and products Diestler obtained convergence of the reaction probability for collision energies up to 1 eV above the threshold. The most important conclusion was that the reaction probability increased rapidly to almost unity when the collision energy was about 0.1 eV above the barrier height, and that the probability for reaction from the $v = 0$ state dipped quite sharply close to the energy at which the $v = 1$ state of the product became open. Schatz and Kupperman†† showed later that this dip also occurred, although rather less sharply, in two- and three-dimensional calculations, and was due to the very high probability of the system being trapped in the symmetric stretching mode of the linear H_3 transition state. This is sometimes called a Feshbach resonance, a term used to describe a situation in which the collision energy is trapped in the internal modes of the system during the collision.

More extensive calculations on the collinear $H + H_2$ reaction were made by

† D. J. Diestler, *J. Chem. Phys.*, **54**, 4547 (1971).

†† G. C. Schatz and A. Kupperman, *J. Chem. Phys.*, **62**, 2502 (1975); **65**, 4668 (1976).

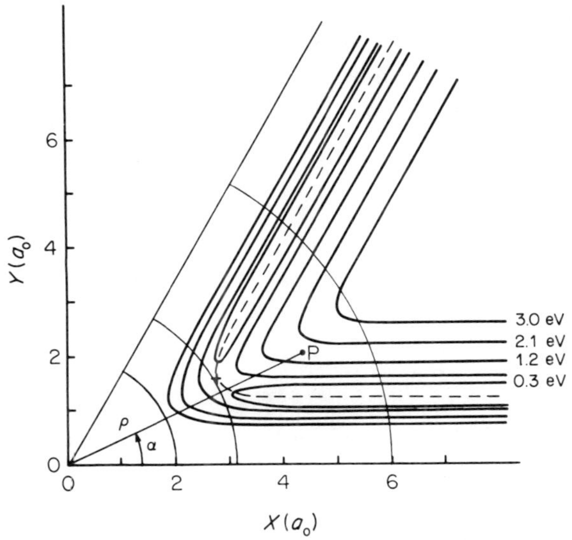

Figure 5.13 Contours of the potential energy surface for collinear $H + H_2$ plotted in Delves scaled coordinates. The arcs at 2, 3.13 and $6a_0$ are used to construct the cross-sections of Figure 5.14 (*After A. Kupperman, J. A. Kaye and J. D. Dwyer*, Chem. Phys. Lett., **74**, *257, 1980*)

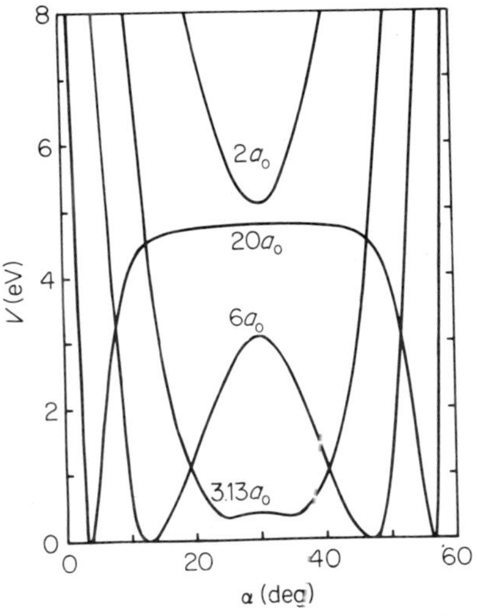

Figure 5.14 Cross-sections $V(\alpha; \rho)$ of the potential shown in Figure 5.13 for selected values of ρ

Kuppermann and coworkers† using the hyperspherical coordinates referred to in Section 5.1. Figure 5.13 shows contours of the H_3 potential plotted in skewed coordinates and Figure 5.14 shows the cross-sections $V(\alpha; \rho)$ for various fixed values of ρ. The hamiltonian in these hyperspherical or scaled Delves coordinates is

$$H(\alpha, \rho) = \frac{-\hbar^2}{2\mu}\left(\frac{\partial^2}{\partial \rho^2} + \frac{1}{\rho}\frac{\partial}{\partial \rho} + \frac{1}{\rho^2}\frac{\partial^2}{\partial \alpha^2}\right) + V(\alpha, \rho) \qquad 5.36$$

In setting up the multichannel equations we could use an adiabatic basis [cf. (3.54)], choosing eigenfunctions $\Lambda(\alpha; \rho)$ of

$$H^0(\alpha; \rho) = \frac{-\hbar^2}{2\mu\rho^2}\frac{\partial^2}{\partial \alpha^2} + V(\alpha; \rho) \qquad 5.37$$

where ρ is a parameter. However, this procedure is quite tedious as it is necessary to solve an eigenfunction problem for every value of ρ. A diabatic basis [cf. (3.47)] is normaly taken as the set of eigenfunctions of H^0 when the reaction coordinate is infinite. However, in hyperspherical coordinates these functions are unsuitable because they will span, and give unnecessary importance to, the infinite space of the H + H + H plateau; the important region of the $H + H_2$ valleys occurs in an infinitesimal range of α. Kuppermann and coworkers suggested as a diabatic basis the eigenfunctions for a fixed finite value of ρ, and chose the value $6a_0$. These eigenfunctions are shown in Figure 5.15. Because V is infinite at $\alpha = 0$ and $\alpha = 60°$ (the skew angle) these eigenfunctions are an infinite but discrete set.

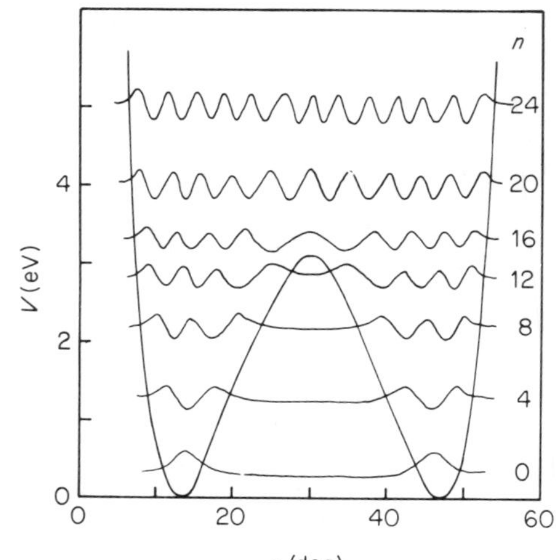

Figure 5.15 Eigenfunctions of the diabatic basis obtained from $V(\alpha; 6a_0)$. Only the even functions (symmetric to reflection in $\alpha = 30°$) are shown

†A. Kuppermann, J. A. Kaye and J. D. Dwyer, *Chem. Phys. Lett.*, **74**, 257 (1980).

Expanding the complete wave functions in the diabatic eigenfunctions $\chi_n(\alpha; \bar{\rho})$ [eigenvalues $E_n(\rho)$]

$$\psi(\alpha, \rho) = \rho^{-1/2} \sum_{n=0}^{N} g_n(\rho; \bar{\rho}) \chi_n(\alpha; \bar{\rho}) \tag{5.38}$$

leads to multichannel equations, which in matrix form are

$$\frac{-\hbar^2}{2\mu} \frac{\partial^2 \mathbf{g}}{\partial \rho^2} + \mathbf{W}\mathbf{g} = \mathbf{E}\mathbf{g} \tag{5.39}$$

where $\mathbf{W}$ is the potential energy interaction matrix whose elements are

$$W_{nn'}(\rho; \bar{\rho}) = \langle n | V(\alpha, \rho) - (\bar{\rho}/\rho)^2 V(\alpha, \bar{\rho}) | n' \rangle \tag{5.40}$$

$\mathbf{E}$ is a diagonal energy matrix with elements

$$E_{nn'}(\rho; \bar{\rho}) = \left[E + \left(\frac{\hbar^2}{8\mu\rho^2} \right) - E_n(\bar{\rho}) \right] \delta_{nn'} \tag{5.41}$$

These equations are integrated numerically from $\rho = 0$ to the asymptotic region, just as for inelastic scattering, to produce a set of independent functions which are then matched to the bound-state eigenfunctions of the reactant and product diatomic molecules to produce scattering matrices. Kuppermann and coworkers found that for a collision energy of up to 1.03 eV (at this energy only two channels are open) only four channels are needed to converge the reaction probability between the ground vibrational states of reactants and products to better than 0.02.

Figure 5.16 shows the reaction probabilities P^{R}_{nn} associated with the process

$$\mathrm{H} + \mathrm{H}_2(n) = \mathrm{H}_2(n) + \mathrm{H} \tag{5.42}$$

These are elastic reaction probabilities—often called vibrationally adiabatic because the vibrational state of the reactant and product H_2 is the same. It is interesting that these functions are all rather similar, and this implies that a good model of the dynamics is to assume the reaction occurs on a one-dimensional potential $V_n(r)$ along the reaction coordinate (r) whose value is the potential along r plus the vibrational energy in state n for motion orthogonal to this coordinate. This is the underlying concept of the Marcus natural coordinate system described in Section 5.1.

In Figure 5.17 are shown the probabilities for the inelastic reaction process P^{R}_{01}. When the collision energy is sufficient to open up the $v = 1$ channel we get a rapid increase in this probability and a decrease in P^{R}_{00}, as already mentioned. Although P^{R}_{01} is generally smaller than P^{R}_{00} it is not so much smaller that one can say the $\mathrm{H} + \mathrm{H}_2$ reaction is predominantly vibrationally adiabatic.

For collisions in three dimensions the most commonly used techniques are either to use different coordinates and basis sets in different reaction channels and to match wave functions at some surface in the interaction region to establish

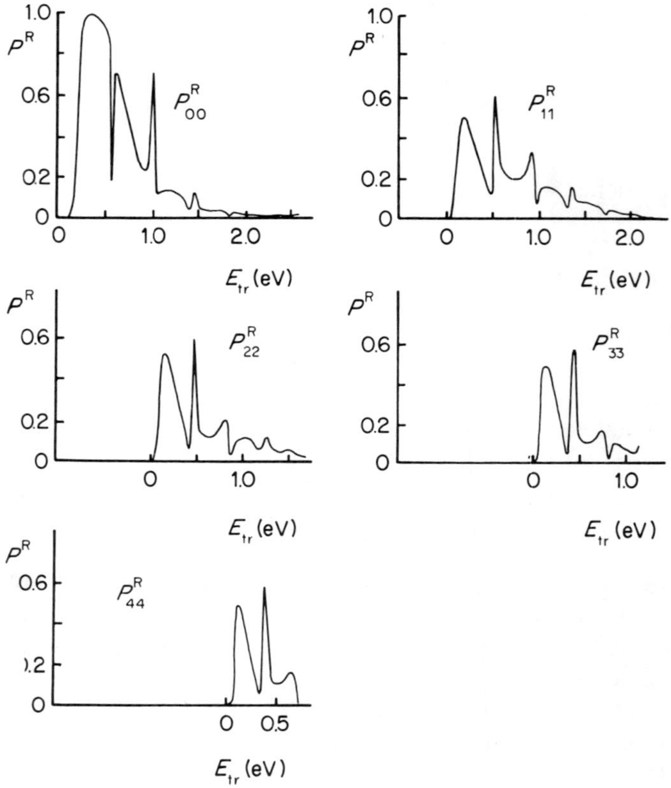

Figure 5.16 Reaction probabilities P^R_{nn} for the potential of Figure 5.13 as a function of the collision translational energy. Superimposing the figures gives a common scale of initial total energy. (*After A. Kuppermann, J. A. Kaye and J. D. Dwyer*, Chem. Phys. Lett., **74**, *257, 1980*)

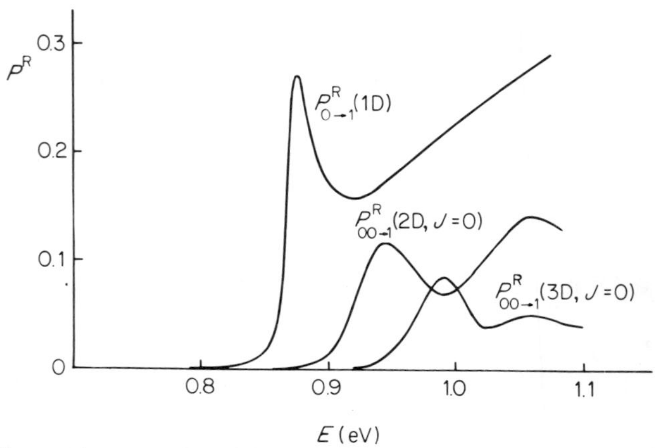

Figure 5.17 Comparison of one-dimensional, two-dimensional and three-dimensional reaction probabilities for the $H + H_2$ exchange reaction $H + H_2(v=0) \rightarrow H_2(v=1) + H$ as a function of the total energy, relative to $H + H_2$ ($v=0$, $j=0$)

continuity in ψ and its derivatives, or to use the generalized hyperspherical coordinates and their associated basis functions.

As in the case of three-dimensional inelastic scattering the rotational angular momentum of the fragments and their orbital angular momentum about the centre of mass will couple to give a total angular momentum J which is conserved. In other words, as for inelastic scattering the coupled equations can be solved for each value of J and partial cross-sections obtained for each J. The total cross-sections will then be given by an expression similar to (4.75).

In addition, if any of the atoms in the colliding molecules are identical, as they are for $H + H_2$, then a symmetry condition must be imposed on the total wave function which will further decouple the equations. For example, in an A + BC rearrangement collision, if atoms A and C are identical, then the scattering wave function will either be symmetric or antisymmetric to the exchange of A and C and the coupled equations can be solved for these separately. Moreover, one of these sets involves only even rotational states of the homonuclear diatomic AC and the other only odd states. More details of the general multichannel theory for the collision of an atom and a diatomic molecule with rearrangement, and with equivalent atoms, can be found in an article my Miller.†

Wolken and Karplus†† were the first to solve the coupled equations for the $H + H_2$ reaction in three dimensions using a small rotational basis ($j = 0$ to 6), only the ground vibrational states, and values of the total angular momentum up to 12. Their main conclusion from this work was that the threshold for reaction was significantly lower than that obtained from classical trajectories on the same surface, illustrating the importance of tunnelling at the threshold of reactions.

The simple view of tunnelling is that it is the passage of particles into regions where the potential energy is greater than the total energy; in consequence the momentum in such regions is imaginary. However, of equal importance is the zero-point energy of the system, because this changes as the system proceeds along the reaction coordinate. Figure 5.18 shows the quantum mechanical cross-sections calculated for the $H + H_2$ exchange reaction (the reactant H_2 being in its lowest quantum state) in the collinear, coplanar and three-dimensional models. The threshold energy increases by about 0.06 eV with each additional dimension, and this is approximately equal to the zero-point energy for the bending mode (0.06 eV in two dimensions and 0.12 eV in three dimensions) in the linear H_3 transition state.

We complete this section by mentioning that the difficulties of applying quantum theory to reactive scattering were first recognized in nuclear physics and in consequence an alternative approach was developed using the momentum representation of wave functions. In so-called formal scattering theory this led to what are known as the Fadeev equations. However, as yet no method has been found to obtain solutions of these equations for the typically rather complicated

†W. H. Miller, *J. Chem. Phys.*, **50**, 407 (1969).
††G. Wolken and M. Karplus, *J. Chem. Phys.*, **60**, 351 (1974).

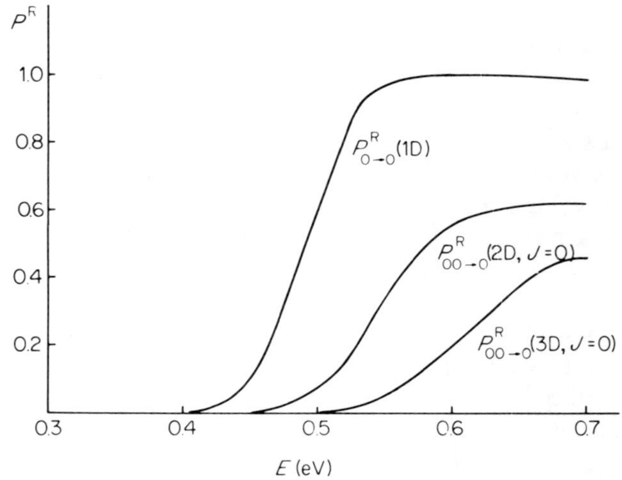

Figure 5.18 Comparison of one-dimensional, two-dimensional and three-dimensional 0—0 reaction probabilities as a function of the total energy

potentials that occur in molecular problems, so that currently all practical problems are being tackled by the standard multichannel methods we have described. The numerical problems are formidable but rapid progress is being made.

5.5 Statistical theories

We have seen in Section 5.3 that the rate constants of chemical reactions in the gas phase can be calculated by taking suitable averages of cross-sections which are functions of the collision energy and the internal states of the reactants and products. In the process of averaging one loses many of the finer features of the crossed beam experiment. Some of these may be associated with the interference features of quantum mechanics; others, such as the dependence of a reaction probability on the internal states of the reactants, are contained also in the classical trajectory method.

If one is only interested in rate constants then they can be approached more directly through one of the statistical theories of chemical reactions: activated complex theory for bimolecular reactions or RRKM theory for unimolecular reactions. It is outside the scope of this book to go into the details of these but we wish to emphasize their undoubted success when used in their most advanced forms. For example, in the variational transition-state theory† of bimolecular reactions, a rate constant is calculated by the standard transition-state formula

†D. A. Truhlar and B. C. Garrett, *Acc. Chem. Res.*, **13**, 440 (1980).

$$k(T) = \frac{kT}{h}\frac{Q^{\neq}(T)}{Q^{R}(T)}\exp\left(-\frac{\Delta V^{\neq}}{kT}\right) \qquad 5.43$$

where $Q^{\neq}$ and Q^{R} are the partition functions of the transition state and reactants respectively and $\Delta V^{\neq}$ is the increase in potential energy along the minimum energy reaction path from the reactants to the transition state. Both $Q^{\neq}$ and $\Delta V^{\neq}$ can be calculated as a function of the position of the transition state along the reaction coordinate, and $k(T)$ can then be minimized with respect to this position.

The basis of this theory is the Wigner definition of the transition state as a dividing surface of the potential energy function such that all classical trajectories starting at the reactants that pass through the transition state go on to form products. In this definition the transition state does not necessarily contain the saddle point on the reaction surface. If 'returning' trajectories are counted as contributing to reaction, then the rate constant will clearly be overestimated; hence a minimization procedure should give the best estimate of the rate constant.

A similar variational procedure can be applied in the RRKM (Rice, Ramsberger, Kassel and Marcus) theory of unimolecular reactions.[†] In this theory the rate constant is given by

$$k(E, J) = \frac{G^{\neq}(E, J)}{h\rho(E, J)} \qquad 5.44$$

where $G^{\neq}$ is the number of energetically available internal states of the transition state critical configuration, and ρ is the density of states of the reactants, both being calculated as a function of the energy and total angular momentum of the reacting molecule. This expression can also be minimized with respect to the position along the reaction coordinate of the transition state. The main emphasis is on methods for the accurate calculation of $G^{\neq}$ and ρ.

A statistical approach which is more relevant to the crossed beam experiment is the information theory, so-called surprisal analysis which was first suggested by Bernstein and Levine.[††] This can applied to inelastic, or more usually, reactive cross-sections, with the aim of predicting *a priori* distributions for the internal energies of the products and interpreting departures from this.

An underlying principle of statistical thermodynamics is that unless there is information to the contrary all sets of states consistent with a given total energy and other external constraints are equally probable. The prior distribution is obtained by applying this principle to the internal energy states of the products of a molecular collision.

[†] W. L. Hase, *Acc. Chem. Res.*, **16**, 258 (1983).
[††] R. D. Levine and R. B. Bernstein, *Acc. Chem. Res.*, **7**, 393 (1974).

The density of translational states per unit volume and unit interval of energy is

$$\rho_T(E) = \left(\frac{\mu^3 E}{2\pi^4 \hbar^6} \right)^{1/2} \qquad 5.45$$

If the products of the collision are AB + C, the internal ro–vibrational states of AB, with energy E_{vj}, have a degeneracy of $2j + 1$, so that the total density of states at an energy E is

$$\rho(E) = \sum_{v=0}^{v_{max}} \sum_{j=0}^{j_{max(v)}} (2j + 1)\rho_T(E - E_{vj}) \qquad 5.46$$

where v_{max} and $j_{max(v)}$ are the largest energetically allowed vibrational and rotational quantum numbers respectively, consistent with the maximum energy E.

The prior distribution of ro–vibrational states of AB will then be

$$P^0(v, j) = \frac{(2j + 1)\rho_T(E - E_{vj})}{\rho(E)} \qquad 5.47$$

and

$$P^0(v) = \sum_{j=0}^{j_{max(v)}} P^0(v, j) \qquad 5.48$$

will be the prior distribution of vibrational states for any j.

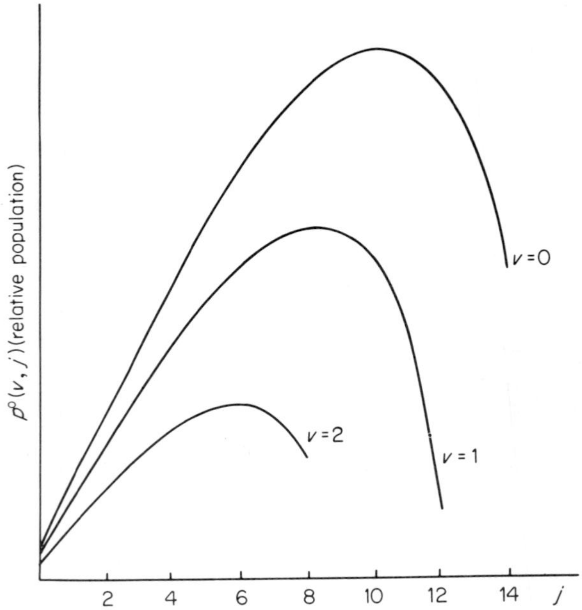

Figure 5.19 Prior distributions for the populations of HD(v, j) produced in reaction (5.49) at a total energy of 1.51 eV. $v = 3$, $j = 0$ is just open at this energy but is negligibly populated

Figure 5.19 shows the prior distributions for the populations of the product states in the reaction

$$H + D_2(0,0) \rightarrow HD(v, j) + D \qquad 5.49$$

at a total energy of 1.51 eV. The vibrational priors, calculated from expression (5.48), are 0.57, 0.31 and 0.12 for $v = 0$, 1 and 2 respectively. The reaction has been studied experimentally[†] at this energy. There is a strong preference for populating the ground vibrational state of HD although accurate vibrational populations have not been determined because, for experimental reasons, not all ro–vibrational states have been detected. The most populated rotational levels for each observed vibrational state are in the range 2–6, which is significantly lower than the maxima in the prior distributions. In fact both the $v = 1$ and $v = 2$ rotational distributions are approximately Boltzmann, with effective temperatures of 1620 and 810 K respectively.

Differences between the prior distribution and an experimental distribution (or even distributions computed by classical trajectory or quantum close-coupled calculations) can be measured by the surprisal, which is defined by

$$I(v, j) = -\ln\left(\frac{P(v, j)}{P^0(v, j)}\right) \qquad 5.50$$

where $P(v, j)$ is the 'observed' distribution. A vibrational surprisal can be similarly defined from the ratio of $P^0(v)$ and $P(v)$.

It is quite common for systems to show a linear surprisal, which is one that can be fitted to the formula

$$I(v, j) = \lambda_0 + \lambda_V f_V + \theta_R g_R \qquad 5.51$$

where f and g are reduced parameters, being the fractions of available energy in rotation and vibration respectively:

$$f_V = \frac{E_V}{E}, \quad g_R = \frac{E_j}{E - E_V} \qquad 5.52$$

The rotational distributions for reaction (5.49) show linear surprisal, as also do the vibrational distributions determined from classical trajectory calculations. The assumption of linear surprisal, with parameters λ_v and θ_R determined from a few experimentally determined populations, allows one to predict populations $P(v, j)$ for unobserved levels. However, there are many cases where populations do not follow a linear surprisal. Although there have been proposals regarding the physical significance of linear and non-linear surprisal, there is no widely accepted view.

[†]D. P. Gerrity and J. J. Valentini, *J. Chem. Phys.*, **79**, 5202 (1983); **81**, 1298 (1984). E. E. Marinero, C. T. Rettner and R. N. Zare, *J. Chem. Phys.*, **80**, 4142 (1984).

Chapter 6

Electronic transitions

6.1 Beyond the Born–Oppenheimer approximation

At high collision energies transitions frequently occur to several electronic states of the collision partners or, in the case of chemical reaction, to several electronic states of the products. Ionization and electron transfer can also occur. In fact one does not always need high collision energies to observe these electronically inelastic events because there are many systems where the potential surfaces of different electronic surfaces come close together. This is particularly so when we are dealing with the collisions of molecules in excited states, or where there are states whose energies differ only through spin–orbit coupling.

Electronically inelastic events involve a transfer of the nuclear motion from one potential energy surface to another; indeed if a large number of electronic states are involved, the concept of nuclear motion on a potential surface may have little relevance to the situation. Transitions between potential energy surfaces are called non-adiabatic processes, and must be described by a theory that goes beyond the Born–Oppenheimer approximation.

Electrons, being much lighter than nuclei, adjust their spatial distribution almost instantaneously to movements of the nuclei. The electronic wave functions and energies are therefore functions of the nuclear positions but usually not, to a first approximation, functions of the nuclear momenta. To go beyond the Born–Oppenheimer approximation we must include the coupling of electronic states by the nuclear momenta operators.

A characteristic time for distinguishing electronic states i and j is $\tau_{ij} = h/\Delta E_{ij}$. For a 1 eV separation this is about 4×10^{-15} s. If the collision time is much greater than τ_{ij}, the dynamics will be predominantly adiabatic, occurring on surfaces i or j.[†] If the collision time is much less than τ_{ij}, the dynamics will be predominantly non-adiabatic, the adiabatic surfaces having no special significance for the nuclear motion. For example, 10 eV hydrogen atoms, interacting

[†]The ratio of these times is proportional to the so-called Massey parameter

$$\gamma = l\Delta E_{ij}/\hbar v$$

where l is the length of the interaction region which is crossed with a velocity v.

over a range of 4 Å have a collision time of $\sim 10^{-14}$ s so that, even at such high collision energies (by thermal energy standards), non-adiabatic processes will have a low probability for electronic surfaces separated by more than 1 eV.

Electronic wave functions are normally expressed in coordinates (**q**) referred to the nuclear framework; the nuclear coordinates (**Q**) are referred to a nuclear centre-of-mass system. The fact that the true centre of mass also depends on the electrons can be ignored for most purposes. The hamiltonian for the whole system in the centre-of-mass system can be written

$$H = T_n(\mathbf{Q}) + T_e(\mathbf{q}) + V(\mathbf{q}, \mathbf{Q}) \qquad 6.1$$

where T_n and T_e are the nuclear and electronic kinetic energy operators respectively:

$$T_n = -\sum_k \frac{\hbar^2}{2\mu_k} \frac{\partial^2}{\partial Q_k^2}, \qquad T_e = -\sum_i \frac{\hbar^2}{2m} \frac{\partial^2}{\partial q_i^2} \qquad 6.2$$

where μ_k are reduced masses associated with nuclei coordinates Q_k and m is the electron mass.

In the Born–Oppenheimer approximation the electronic states are eigenfunctions of the electronic hamiltonian, which is the full hamiltonian (6.1) less the nuclear kinetic energy operator

$$H_e(\mathbf{q}, \mathbf{Q}) = H - T_n(\mathbf{Q}) \qquad 6.3$$

$$[H_e(\mathbf{q}, \mathbf{Q}) - V_j(\mathbf{Q})]\Psi_j(\mathbf{q}; \mathbf{Q}) = 0 \qquad 6.4$$

where $\Psi_j(q; \mathbf{Q})$ are the wave functions of the adiabatic electronic states and their energies $V_j(\mathbf{Q})$ are the adiabatic potentials; both are parametrically dependent on the nuclear coordinates.

Eigenfunctions of the full hamiltonian can be obtained as an expansion in products of the adiabatic functions and nuclear functions $\Phi_j(\mathbf{Q})$ as follows:

$$[T_n(\mathbf{Q}) + H_e(\mathbf{q}, \mathbf{Q}) - E]\sum_j \Psi_j(\mathbf{q}; \mathbf{Q})\Phi_j(\mathbf{Q}) = 0 \qquad 6.5$$

Multiplying this from the left by a function $\Psi_i(\mathbf{q}; \mathbf{Q})$, integrating over the space of **q** and using (6.4), we obtain

$$\sum_j [\langle \Psi_i(\mathbf{q}; \mathbf{Q}) | T_n(Q) | \Psi_j(\mathbf{q}; \mathbf{Q})\Phi_j(\mathbf{Q})\rangle + \delta_{ij}(V_j(\mathbf{Q}) - E)\Phi_j(\mathbf{Q})] = 0 \qquad 6.6$$

The adiabatic nuclear wave functions are derived by assuming that the $\Psi_j(\mathbf{q}; \mathbf{Q})$ are slowly varying functions of **Q**, so that terms like $\partial\Psi_j(\mathbf{q}; \mathbf{Q})/\partial Q_k$ can be neglected. In this case we can write

$$T_n(\mathbf{Q})\Psi_j(\mathbf{q}; \mathbf{Q})\Phi_j(\mathbf{Q}) = \Psi_j(\mathbf{q}; \mathbf{Q})T_n(\mathbf{Q})\Phi_j(\mathbf{Q}) \qquad 6.7$$

and the nuclear functions satisfy the equations

$$[T_n(\mathbf{Q}) + V_i(\mathbf{Q}) - E]\Phi_i(\mathbf{Q}) = 0 \qquad 6.8$$

There are many situations where the electronic wave functions are not slowly varying functions of the nuclear coordinates so that approximation (6.7) cannot be made. Figure 6.1 illustrates such a case, typical of all alkali halides. The lowest energy dissociation products of LiF are the neutral atoms, the ions being higher in energy by the difference between the ionization potential of Li and the electron affinity of F. The interaction of the neutral atoms is weak at large R because it depends on the overlap of Li and F orbitals. In contrast the ions have a strong long-range coulomb attractive energy, $-e^2/R$, so that when

$$\frac{e^2}{R} > I_{\text{Li}} - A_{\text{F}} \qquad 6.9$$

the ionic component has the lower energy. For LiF this occurs at $R \sim 7$ Å, and in this region the ground-state wave function is rapidly changing its character—

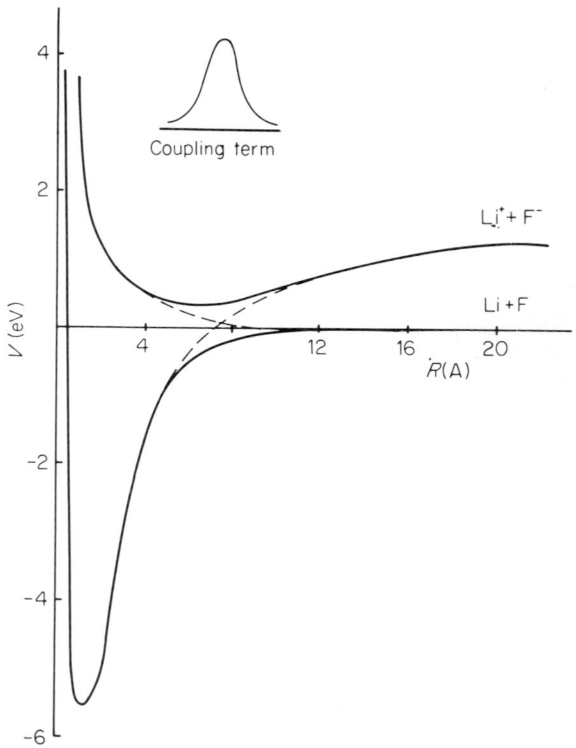

Figure 6.1 Ionic and covalent diabatic curves (dashed) and adiabatic curves (full) for LiF. The diabatic crossing point is at approximately 7.4 Å. The coupling term in the adiabatic basis (not to scale) is a lorentzian function (see equation (6.46))

from covalent at large R to ionic at small R. If Li and F approach one another and get closer than 7 Å, then adiabatically an electron jumps from Li to F. As they move away the electron will jump back at about 7 Å. However, if the electron jumps on the way in but not on the way out, or vice versa, then the reactants Li and F become products Li^+ and F^-; this is called a non-adiabatic process. Non-adiabatic processes are very common when there is avoided crossing, as illustrated in Figure 6.1, but they are not confined to these situations.

The intersection of ionic and covalent potentials is very common when one of the collision partners has a low ionization potential and the other a low electron affinity. As the ionic potential is strongly attractive, the process of electron transfer can lead to a low energy path for reaction, often without an energy barrier, a process which Polanyi called harpooning. The reaction of Li and F_2 to give LiF + F is one example; the reaction is very rapid with a cross-section of the order of 100 Å^2.

To calculate a cross-section or transition probability for a non-adiabatic process from adiabatic wave functions, we have to calculate matrix elements of $\partial/\partial Q_k$ and $\partial^2/\partial Q_k^2$. These are large, for example, in the regions of avoided crossing, as illustrated in Figure 6.1. However, these matrix elements are not easy to calculate even if the wave functions are known, and in many cases they are not known. An alternative approach is to find another set of electronic functions, called the diabatic basis, for which such matrix elements can be taken as zero. In this case, the non-adiabatic processes arise from the coupling of diabatic states by potential energy terms.

Let us assume that an electronic basis can be found, $\Psi'(\mathbf{q};\mathbf{Q})$, for which all derivatives of $\partial/\partial Q_k$ and $\partial^2/\partial Q_k^2$ can be neglected. If this is a complete set of states it can be used to expand the adiabatic functions

$$\Psi_j(\mathbf{q};\mathbf{Q}) = \sum_p C_{jp}(\mathbf{Q})\Psi'_p(\mathbf{q};\mathbf{Q}) \qquad 6.10$$

Now it might appear that the derivatives of Ψ' can only be neglected if the basis is defined with respect to a fixed nuclear geometry $\mathbf{Q}_0$, but in practice such a set would give a poor representation of the adiabatic states far from $\mathbf{Q}_0$ unless a very large number of terms was included in the expansion. The condition we require is much less severe, and that is that the derivatives of $\Psi'(\mathbf{q};\mathbf{Q})$ can be neglected in comparison with the derivatives of $\mathbf{C}(\mathbf{Q})$.

We can transform the set of equations (6.6) to the new basis $\Psi'(\mathbf{q};\mathbf{Q})$ using (6.10). The transformation is most easily accomplished by expressing (6.6) in matrix form and to do that we must use the full expression for T_n, which is (6.2), and write (6.6) as

$$-\sum_k \frac{\hbar^2}{2\mu_k}\{\langle\partial^2/\partial Q_k^2\rangle\boldsymbol{\phi}(\mathbf{Q}) + 2\langle\partial/\partial Q_k\rangle[\partial\boldsymbol{\phi}(\mathbf{Q})/\partial Q_k] + [\partial^2\boldsymbol{\phi}(\mathbf{Q})/\partial Q_k^2]\}$$
$$+(\mathbf{V}-E\mathbf{I})\boldsymbol{\phi}(\mathbf{Q}) = \mathbf{0} \qquad 6.11$$

where $\langle\ \rangle$ are square matrices for the adiabatic basis, $\boldsymbol{\phi}(\mathbf{Q})$ is a column matrix and $\mathbf{V}$ is a diagonal matrix.

Writing

$$\boldsymbol{\Psi}(\mathbf{q};\mathbf{Q}) = \mathbf{C}(\mathbf{Q})\boldsymbol{\Psi}'(\mathbf{q};\mathbf{Q}) \tag{6.12}$$

and neglecting derivatives of $\boldsymbol{\Psi}'$ leads to

$$-\sum_k \frac{\hbar^2}{2\mu_k}\{\mathbf{C}(\partial^2\mathbf{C}^\dagger/\partial Q_k^2)\boldsymbol{\phi} + 2\mathbf{C}(\partial\mathbf{C}^\dagger/\partial Q_k)(\partial\boldsymbol{\phi}/\partial Q_k) + (\partial^2\boldsymbol{\phi}/\partial Q_k^2)\} + (\mathbf{V} - E\mathbf{I})\boldsymbol{\phi} = \mathbf{0} \tag{6.13}$$

which can be contracted to

$$\mathbf{C}T_{\mathrm{n}}\mathbf{C}^\dagger\boldsymbol{\phi} + (\mathbf{V} - E\mathbf{I})\boldsymbol{\phi} = \mathbf{0} \tag{6.14}$$

Multiplying from the left by $\mathbf{C}^\dagger$ and noting $\mathbf{C}^\dagger\mathbf{C} = \mathbf{I}$ leads to

$$T_{\mathrm{n}}(\mathbf{C}^\dagger\boldsymbol{\phi}) + (\mathbf{C}^\dagger\mathbf{V}\mathbf{C} - E\mathbf{I})(\mathbf{C}^\dagger\boldsymbol{\phi}) = \mathbf{0} \tag{6.15}$$

or

$$T_{\mathrm{n}}\boldsymbol{\phi}' + (\mathbf{V}' - E\mathbf{I})\boldsymbol{\phi}' = \mathbf{0} \tag{6.16}$$

Thus the nuclear functions in the diabatic basis are related to those in the adiabatic basis by

$$\boldsymbol{\phi}' = \mathbf{C}^\dagger\boldsymbol{\phi} \tag{6.17}$$

and

$$\mathbf{V}' = \mathbf{C}^\dagger\mathbf{V}\mathbf{C} \tag{6.18}$$

is a non-diagonal potential energy matrix. The diagonal elements of $\mathbf{V}'$ are the diabatic potentials; they would, for example, follow the dashed curves in Figure 6.1. Transitions between diabatic surfaces are caused by the off-diagonal elements of $\mathbf{V}'$.

We have not, of course, described methods for constructing the diabatic potential matrix $\mathbf{V}'$. Recipes have been proposed for atom–atom and atom–diatomic molecule collisions and these have been reviewed by Baer.[††]

The most important non-adiabatic processes occur when there is an avoided crossing of two adiabatic surfaces, and in this case it is not difficult to deduce reasonable diabatic potentials by interpolating across the avoided crossing region as in Figure 6.1. Considering just these two states in isolation from the others, the adiabatic potentials will be the eigenvalues of the diabatic potential matrix, which are

$$V_1, V_2 = \tfrac{1}{2}\{V'_{11} + V'_{22} \pm [(V'_{11} - V'_{22})^2 + 4V'^2_{12}]^{1/2}\} \tag{6.19}$$

Thus knowing V_1 and V_2 and interpolating V'_{11} (say) is sufficient to dermine V'_{22} and V'_{12}.

To calculate the cross-sections for non-adiabatic processes we therefore have to calculate the nuclear momentum coupling terms in the adiabatic basis or the potential coupling terms in a diabatic basis (which is generally simpler), and this must be done for all nuclear variables which are $3N - 3$ in number relative to a

[††] M. Baer, in J. M. Bowman (ed.), *Molecular Collision Dynamics*, Ch. 5, Springer-Verlag, 1983.

fixed centre of mass. Three of these variables describe rotation of the whole system (two for atom–atom scattering) and the remainder are for the collision or radial coordinate, and for the vibrational motion or relative rotational motions of the colliding species.

Broadly speaking, this leads to a division of coupling terms into 'angular' and 'radial', the former being of most interest in spectroscopy and the latter in scattering theory. The distinction between the two depends on the nature of the scattering fragments and we will therefore turn to an examination of the simplest cases in more detail. It is, however, worth emphasizing that the treatment of non-adiabatic effects in scattering is a more complicated problem than in spectroscopy. In spectroscopy the internuclear distances generally cover a rather narrow range, and there is a 'best' coupling scheme for the angular momentum (nuclear, orbital and spin) described succinctly by the so-called Hund coupling cases. In scattering theory we are interested in a range of internuclear distances going from infinity to the classical turning points, and over this range the 'best' coupling case may change. The manner in which spin–orbit coupling, for example, changes with internuclear distance will be relevant to understanding the populations of, say, $F(^2P_{1/2})$ and $F(^2P_{3/2})$ in a collision that produces fluorine atoms.

6.2 Atom–atom collisions

The non-adiabatic collisions of atoms were first examined in detail by Thorson[†] although there had been earlier studies of the problem, particularly in the field of spectroscopy, by Kronig. In a body-fixed coordinate system for the nuclear motion (z is the internuclear axis) the hamiltonian for a diatomic spin-free system is

$$H(\mathbf{q}, r) = H_e(\mathbf{q}, r) - \frac{\hbar^2}{2\mu r^2}\frac{\partial}{\partial r}\left(r^2 \frac{\partial}{\partial r}\right) + \frac{1}{2\mu r^2}(\mathbf{J}^2 + \mathbf{L}^2 - 2\mathbf{J}_z\mathbf{L}_z - \mathbf{J}_+\mathbf{L}_- - \mathbf{J}_-\mathbf{L}_+) \quad 6.20$$

where $\mathbf{J}$ is the total and $\mathbf{L}$ the electronic orbital angular momentum operators, with

$$\mathbf{J}_\pm = \mathbf{J}_x \pm \mathrm{i}\mathbf{J}_y, \qquad \text{etc.} \quad 6.21$$

If the system has non-zero electron spin then $\mathbf{L}$ must be replaced by $\mathbf{L} + \mathbf{S}$.

The angular coupling terms arise from the terms in $\mathbf{J}_\pm$ and $\mathbf{L}_\pm$ (or the appropriate operator if spin is present) and is a Coriolis interaction. This is the general term used to describe coupling between motion in a body-fixed frame (e.g. movement on the earth's surface) and the rotational motion of that frame (the earth's rotation). In respect of the non-adiabatic coupling now being considered, the electrons (whose positions are referred to the body-fixed frame of the nuclei) experience a force from the motion of the nuclei; one can say they lag behind the

[†]W. R. Thorson, *J. Chem. Phys.*, **34**, 1744 (1961).

nuclear motion. A typical situation in which these terms need to be invoked is the change in the electronic state of an atom on collision with a proton. For example, the inelastic collision.

$$H^+ + H(1s) \rightarrow \begin{bmatrix} H^+ + H(2s) \\ H^+ + H(2p) \end{bmatrix} \qquad 6.22$$

has been studied extensively at high collision energies and above about 60 keV charge exchange between the protons has been found to be negligible. The excitation processes have been treated by the first Born approximation, the distorted wave approximation and by coupled-channel methods. The Born approximation gives a very poor value for the H(2s) cross-section because this is mainly produced by the indirect route:

$$H(1s) \rightarrow H(2p) \rightarrow H(2s)$$

The second of these steps has a high probability because the 2s and 2p states of hydrogen are dengenerate.

We turn now to a consideration of the radial coupling term in atom–atom collisions with particular reference to the curve-crossing problem illustrated in Figure 6.1; this is typical of situations in which electron transfer can occur on collision between two atoms or ions. This type of problem can be solved by the coupled-channel method using the diabatic or adiabatic representations described earlier. However, analysis of such problems shows that the cross-section is almost entirely determined by the potentials in the crossing region and in consequence most such problems can be treated using just two coupled channels. If there are mutliple crossings and more than two potential curves involved, then these are usually well separated and they can be treated as a series of two-state interactions. This is not to say that it is computationally difficult to carry out such calculations with several channels, but the essential physics can usually be extracted from the two-channel problem.

We will choose to work in a diabatic basis so that the coupling terms are in the potential energy [equation (6.16)]. We will assume that there is no spin or electronic orbital angular momentum; the total angular momentum **J** just consists of the nuclear angular momentum and contributes a centrifugal potential $\hbar^2 l(l+1)/(2\mu r^2)$ for each partial wave. Using the symbols adopted for our treatment of elastic scattering we must solve the two-channel equivalent of (2.36) which is

$$\begin{aligned} \left[\frac{d^2}{dr^2} + k^2 - U_{11}(r) - \frac{l(l+1)}{r^2}\right]\psi_1(r) - U_{12}(r)\psi_2(r) = 0 \\ \left[\frac{d^2}{dr^2} + k^2 - U_{22}(r) - \frac{l(l+1)}{r^2}\right]\psi_2(r) - U_{12}(r)\psi_1(r) = 0 \end{aligned} \qquad 6.23$$

where

$$U_{ij} = \frac{2\mu(V_{ij} + \delta_{ij}E_i^0)}{\hbar^2} \qquad 6.24$$

$\mathbf{V}$ being the diabatic potential defined such that all elements go to zero at $r = \infty$ and E_i^0 are the diatomic dissociation limits. ψ_1 and ψ_2 are clearly functions of l but we have dropped this designation for brevity.

The channel wave numbers k_1 and k_2 will be different if the asymptotic (dissociation) energies of the two potentials $U_{11}(r)$ and $U_{22}(r)$ are different, as they are except in cases of resonant charge exchange (e.g. $H^+ + H \rightarrow H + H^+$). We will assume that both channels are open so that k_1^2 and k_2^2 are both positive.

It is a straightforward matter to solve (6.23) for any value of l if the $\mathbf{U}$-matrix elements are given. Starting the integration from a small value of r where ψ_1 and ψ_2 can be taken as zero, with initial conditions $\psi'_1(0) = 1$, $\psi'_2(0) = 0$, or $\psi'_1(0) = 0$, $\psi'_2(0) = 1$, gives asymptotic wave functions

$$\psi(r) \sim e^{i(\mathbf{k}r - l\pi/2)}\mathbf{A} + e^{-i(\mathbf{k}r - l\pi/2)}\mathbf{B} \qquad 6.25$$

where $\mathbf{k}$ is the diagonal matrix of the channel wave numbers. $\mathbf{A}$ and $\mathbf{B}$ are the 2×2 Jost matrices; these also depend on l. As there are no closed channels in this problem the transformation to give asymptotic functions with unit amplitude in the entrance channels

$$\psi(r) \sim e^{i(\mathbf{k}r - l\pi/2)}\mathbf{I} + e^{-i(\mathbf{k}r - l\pi/2)}\mathbf{B}'' \qquad 6.26$$

requires only that

$$\mathbf{B}'' = \mathbf{B}(\mathbf{A})^{-1} \qquad 6.27$$

and the $\mathbf{S}$-matrix elements for the transition from channel 1 to channel 2 is [cf. (A3.9)]

$$S_{12} = -\left(\frac{k_2}{k_1}\right)^{1/2} B''_{12} \qquad 6.28$$

The scattering amplitude is obtained from the partial wave expansion [cf. (2.51) and (4.38)]

$$f_{12}(\theta) = \frac{1}{2i(k_1 k_2)^{1/2}} \sum_l (2l+1) S^l_{12} P_l(\cos\theta) \qquad 6.29$$

and the cross-section is

$$\sigma_{12}(\theta) = \left(\frac{k_2}{k_1}\right) | f_{12} |^{1/2} \qquad 6.30$$

The S^l_{12} matrix elements are oscillating functions of l, or for a fixed l are oscillating functions of the collision energy. We can understand this behaviour by examining the first distorted wave Born approximation (3.100) which is valid if the coupling potential U_{12} is small. The relevant quantity is the integral

$$\int_0^\infty \psi_1^{(0)}(r) U_{12}(r) \psi_2^{(0)}(r)\, dr \qquad 6.31$$

where $\psi_1^{(0)}$ and $\psi_2^{(0)}$ are the distorted waves in the two channels, i.e. the solutions of (6.23) with $U_{12}(r) = 0$. Figure 6.2 illustrates a typical situation for an $l = 0$

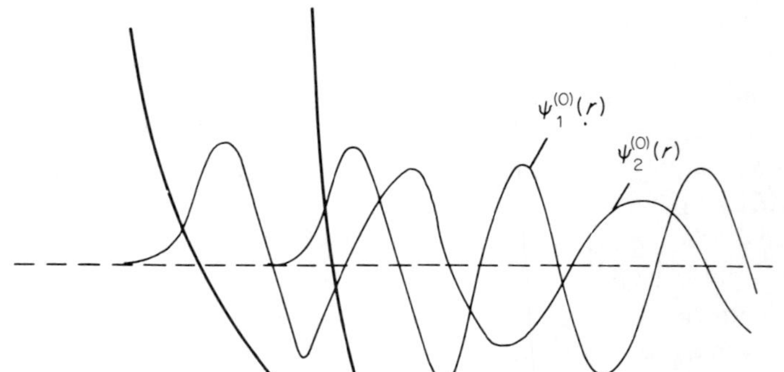

Figure 6.2 Diabatic distorted wave functions which are required for evaluation of the integral (6.31)

partial wave. The distorted waves are oscillating functions with asymptotic limits $\sin(kr + \delta)$. If the two electronic states are related to each other by an electron transfer, one potential being asymptotic to $A^+ + B$ and the other to $A + B^+$, then the interaction potential $U_{12}(r)$ will be related to the overlap of orbitals of A and B and this will decrease exponentially as r increases. The integrand in (6.30) oscillates as the product $\psi_1^{(0)}(r)\psi_2^{(0)}(r)$, and the value of the integral depends on the extent to which these two functions are in phase or out of phase in regions where $U_{12}(r)$ is not negligible; in fact it is the region around the classical turning point of $U_{22}(r)$ that is most important. The integral clearly has a rough proportionality to the quantity

$$\int \psi_1^{(0)}(r)\psi_2^{(0)}(r)\,\mathrm{d}r \qquad 6.32$$

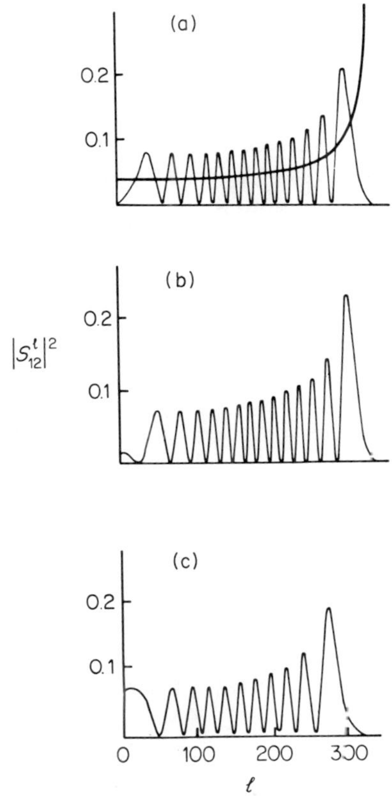

Figure 6.3 Comparison of (a) coupled-channel (b), first-order distorted wave and (c) classical path results for $|S_{12}^{l}|^2$. The potential models He^+ + Ne at collision energy of 70.9 eV. The smooth curve in (a) is the Landau–Zener result

which is called the Franck–Condon integral for the two wave functions.†

Changing the energy will produce different shifts in the phases of the two wave functions; changing l at fixed energy does the same because this has the effect of adding a centrifugal potential to $U_{11}(r)$ and $U_{22}(r)$ and this changes the classical turning points. The crossing point, we note, is independent of l because the centrifugal potential is the same for both electronic states. However, at large values of l the energy at the crossing point will lie above the collision energy and the classical turning points will then be to the right of the crossing point. Further increase in l will therefore lead to a smooth reduction to zero in the value of the integral because $U_{12}(r)$ will become smaller in the region where the two distorted wave functions overlap.

Figure 6.3 shows the values of $|S_{12}^{l}|^2$ that have been calculated for potentials representing the collisional excitation of $Ne(2p^6 \rightarrow 2p^5 3s)$ by He^+ at an energy of

†In bound-state problems the overlap of nuclear wave functions is related to the intensity of vibrational progressions by the Franck–Condon principle.

70.9 eV;[†] the relevant diabatic potentials cross at about $2a_0$. The first-order distorted wave Born approximation and the coupled channel are in very close agreement in this case and the resulting inelastic cross-sections agree well with experimental measurements.

Oscillations in the $\mathbf{S}^l$-matrix elements and in the differential cross-section were first predicted by Stueckelberg for a model two-state curve-crossing problem, which we will now examine in more detail.

The two-state problem for atom–atom collisions has been examined for several model potentials. One of these, called Demkov coupling, or perturbed symmetric resonance, is particularly relevant to the transitions between the multiplet states of open-shell atoms. For example, the ground-state term of a halogen atom is ^2P and this is split by spin–orbit coupling into $^2P_{1/2}$ and $^2P_{3/2}$ states. However, collisions with, say, an inert gas atom will mix these states because in a diatomic molecule only j_z is a conserved quantity; the $m_j = \pm\frac{1}{2}$ components of $^2P_{1/2}$ and $^2P_{3/2}$ will appear in the same block of the hamiltonian matrix. In the Demkov coupling model it is assumed that the two diabatic potentials $U_{11}(r)$ and $U_{22}(r)$ are separated by a small and roughly constant energy and the coupling term $U_{12}(r)$ changes sharply from zero to a value greater than the diabatic separation at some finite value of r. At infinite separation of the two atoms the adiabatic and diabatic states are identical, but when the coupling term becomes large the two adiabatic state wave functions are the sum and difference of the two diabatic state functions. The transition point between these two limiting situations is where

$$|U_{12}(r)| = |U_{11}(r) - U_{22}(r)| \qquad 6.33$$

Another very important model is for diabatic curve crossing as illustrated in Figure 6.1, which is particularly important for charge transfer collisions. This was first treated, independently, by Landau, Zener and Stueckelberg using the potential

$$\begin{aligned} U_{11}(r) - U_{22}(r) &= -(F_1 - F_2)(r - r_c) \\ U_{12}(r) &= U_{12} \text{ (a constant)} \end{aligned} \qquad 6.34$$

where r_c is the crossing point of the diabatic curves. Several methods have been used to solve this problem, time dependent and time independent, and various extensions of the JWKB method. None of these is mathematically trivial. Our derivation is time independent and based on a perturbation treatment of the adiabatic equations.[††] Figure 6.4 illustrates some of the parameters that enter the derivation.

In the so-called Landau–Zener model U_{12} may be large in absolute magnitude, and hence the distorted wave approximation (6.31) will not be valid. However, in this case, the adiabatic curves will be well separated so that to a first approximation the motion occurs on the adiabatic curves. The transition

[†] R. E. Olsen and F. T. Smith, *Phys. Rev.*, **A3**, 1607 (1971). J. P. Braga, L. J. Dunne and J. N. Murrell, *Chem. Phys. Lett.*, **120**, 147 (1985).
[††] S. D. Bosanac, *J. Math. Phys.*, **20**, 396 (1979).

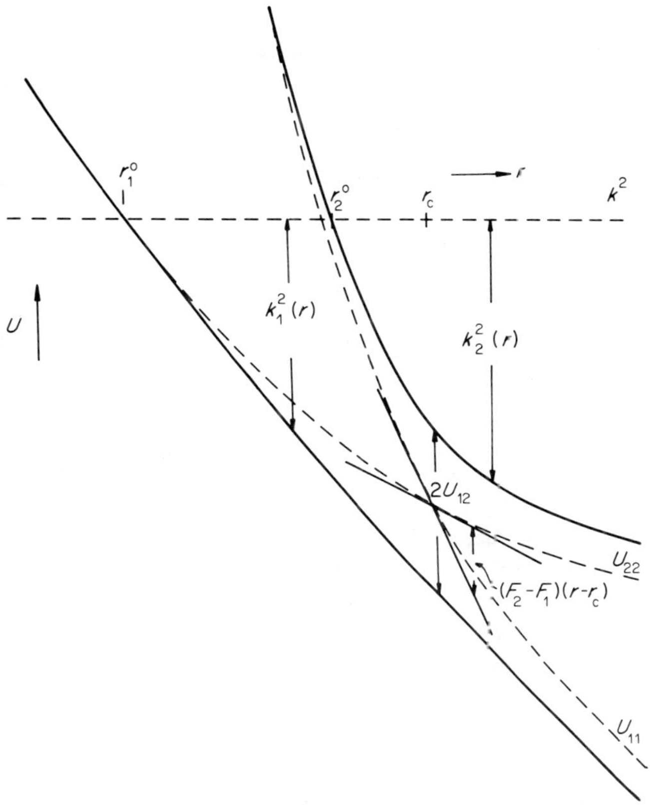

Figure 6.4 The Landau–Zener curve crossing parameters. Diabatic (dashed) and adiabatic (full) curves are shown. F_1 and F_2 are the slopes of the diabatic curves at the crossing point. Other parameters are defined in the text

probability is therefore most easily obtained by working in the adiabatic basis for which the coupling arises from the kinetic energy operator.

Writing

$$\psi = \mathbf{Q}(r)\mathbf{\Phi}(r) \tag{6.35}$$

and multiplying (6.23) from the left by $\mathbf{Q}^\dagger(r)$ leads to the adiabatic equations

$$\left[\mathbf{Q}^\dagger(r)\frac{\mathrm{d}^2}{\mathrm{d}r^2}\mathbf{Q}(r) + \mathbf{k}^2 - \mathbf{\Lambda}(r) - \frac{l(l+1)}{r^2}\mathbf{I}\right]\mathbf{\Phi}(r) = 0 \tag{6.36}$$

where $\mathbf{\Lambda}(r)$ is a diagonal matrix whose elements are the eigenvalues of $\mathbf{U}(r)$.

We write these eigenvalues

$$\bar{U} \pm U_{12}(1+\varepsilon^2)^{1/2} \tag{6.37}$$

where

$$\bar{U} = \tfrac{1}{2}(U_{11} + U_{22}) \tag{6.38}$$

and, from (6.34),

$$\varepsilon = \left|\frac{F_2 - F_1}{2U_{12}}\right|(r - r_c) \qquad 6.39$$

Because the centrifugal term is the same for both diabatic equations, $\mathbf{Q}(r)$ will be independent of l; it is also independent of k^2.

It is convenient to define a matrix $\boldsymbol{\eta}$ which is related to the derivative of $\mathbf{Q}$ by

$$\mathbf{Q}'(r) = \mathbf{Q}(r)\boldsymbol{\eta}(r) \qquad 6.40$$

and because $\mathbf{Q}$ is unitary ($\mathbf{Q}^{\dagger}\mathbf{Q} = \mathbf{I}$), $\boldsymbol{\eta}$ must be antisymmetric.

Expanding the differential in (6.36) and using (6.37) gives the coupled equations in the form

$$(\boldsymbol{\Phi}'' + 2\boldsymbol{\eta}\boldsymbol{\Phi}' + \boldsymbol{\eta}^2\boldsymbol{\Phi} + \boldsymbol{\eta}'\boldsymbol{\Phi}) + \left[\mathbf{k}^2 + \boldsymbol{\Lambda}(r) - \frac{l(l+1)}{r^2}\mathbf{I}\right]\boldsymbol{\Phi} = 0 \qquad 6.41$$

in which the only non-diagonal terms are those containing $\boldsymbol{\eta}$; this is therefore the coupling matrix in the adiabatic basis.

For the two-channel case we can write

$$\boldsymbol{\eta} = a(r)\begin{bmatrix} 0 & 1 \\ -1 & 0 \end{bmatrix} \qquad 6.42$$

where $a(r)$ is a scalar and $\boldsymbol{\eta}^2$ which occurs in (6.41) is equal to $a^2(r)\mathbf{I}$ and does not lead to coupling.

The transformation matrix $\mathbf{Q}(r)$ can be written

$$\mathbf{Q} = \begin{bmatrix} \cos\phi & \sin\phi \\ -\sin\phi & \cos\phi \end{bmatrix} \qquad 6.43$$

where

$$\tan 2\phi(r) = -\frac{2U_{12}}{U_{11} - U_{22}} = \frac{1}{\varepsilon} \qquad 6.44$$

Now

$$\mathbf{Q}' = \begin{bmatrix} -\sin\phi & \cos\phi \\ -\cos\phi & -\sin\phi \end{bmatrix}\boldsymbol{\Phi}' \qquad 6.45$$

and it is then simple to show that

$$a(r) = \Phi'(r) = \frac{F_2 - F_1}{4U_{12}(1 + \varepsilon^2)} \qquad 6.46$$

Note that $a(r)$ is a lorentzian function centred at $r = r_c$, as illustrated in Figure 6.1.

To proceed further we return to equation (6.41) and handle the coupling, which for the two-channel problem is due to the terms $\boldsymbol{\eta}\boldsymbol{\Phi}'$ and $\boldsymbol{\eta}'\boldsymbol{\Phi}$. Neglecting these decouples the equations and we obtain the zeroth-order distorted waves for the adiabatic potentials. By analogy with (6.31) the first-order distorted wave solution will depend on the integral

$$M_{12} = \int_0^\infty \Phi_1^{(0)}\left(2\eta_{12}\frac{\mathrm{d}}{\mathrm{d}r} + \eta'_{12}\right)\Phi_2^{(0)}\,\mathrm{d}r \qquad 6.47$$

and on introducing $a(r)$ from (6.42) and symmetrizing the resulting integral, using integration by parts, we have

$$M_{12} = \int_0^\infty (\Phi_1^{(0)}\Phi_2^{(0)\prime} - \Phi_1^{(0)\prime}\Phi_2^{(0)})a(r)\,\mathrm{d}r \qquad 6.48$$

As $a(r)$ has its maximum value at the crossing point of the diabatic curves, the major contribution to the integral comes from this region. We can now use the JWKB approximation to the distorted waves to evaluate this integral. For a left-hand turning point (2.76),

$$\Phi_i^{(0)}(r) \sim k_i^{-1/2}(r)\sin\left[\int_{r_i^0}^r k_i(r')\,\mathrm{d}r' + \frac{\pi}{4}\right] \qquad 6.49$$

where

$$k_i^2(r) = k^2 - \Lambda_{ii} - \frac{l(l+1)}{r^2} \qquad 6.50$$

and r_i^0 is the classical turning point for the adiabatic effective potential; this is a function of l. Expression (6.49) applies away from the classical turning point; hence to use this in (6.45) it must be assumed that the major contribution to the integral occurs away from *both* classical turning points. In practice this means that the crossing point of the diabatic potentials, r_c, which is where $a(r)$ has its greatest value, must be significantly greater than both r_1^0 and r_2^0.

From (6.49) we find

$$\begin{aligned}\Phi_1^{(0)}\Phi_2^{(0)\prime} - \Phi_1^{(0)\prime}\Phi_2^{(0)} = [k_1(r)k_2(r)]^{-1/2}\Bigg\{k_1(r)\cos\left[\int_{r_1^0}^r k_1(r')\,\mathrm{d}r' + \frac{\pi}{4}\right]\\ \times \sin\left[\int_{r_2^0}^\infty k_2(r')\,\mathrm{d}r' + \frac{\pi}{4}\right] - k_2(r)\sin\left[\int_{r_1^0}^r k_1(r')\,\mathrm{d}r' + \frac{\pi}{4}\right]\\ \times \cos\left[\int_{r_1^0}^r k_2(r')\,\mathrm{d}r' + \frac{\pi}{4}\right]\Bigg\}\end{aligned} \qquad 6.51$$

Since most of the contribution to M_{12} comes from the crossing point, which is where the two adiabatic curves make their closest approach, we can assume that the factors $(k_1/k_2)^{1/2}$ and $(k_2/k_1)^{1/2}$ that appear in (6.51) are equal, and in this case (6.51) reduces to

$$\Phi_1^{(0)}\Phi_2^{(0)\prime} - \Phi_1^{(0)\prime}\Phi_2^{(0)} \approx \sin\left\{\delta^0 + \int_{r_c}^r [k_2(r') - k_1(r')]\,\mathrm{d}r'\right\} \qquad 6.52$$

where

$$\delta^0 = \int_{r_2^0}^{r_c} k_2(r')\,\mathrm{d}r' - \int_{r_1^0}^{r_c} k_1(r')\,\mathrm{d}r' \qquad 6.53$$

is called the Stueckelberg phase; this depends of course on l and on the total energy and is the term mainly responsible for the oscillations of the transition probability with these variables.

If we define an average wave vector for the diabatic potentials by

$$\bar{k}(r) = \left[k^2 - \tfrac{1}{2}(U_{11} + U_{22}) - \frac{l(l+1)}{r^2} \right]^{1/2} \tag{6.54}$$

then taking U_{12} to be smaller than $\bar{k}^2$, we can show that

$$k_2(r) - k_1(r) = \frac{U_{12}(1+\varepsilon^2)^{1/2}}{\bar{k}(r)} \tag{6.55}$$

As the main contribution to M_{12} arises from the region of r_c we can take $k(r)$ to be equal to its value at the crossing point, whence the integral in (6.52) becomes (using standard integrals)

$$\int_{r_c}^{r} [k_2(r') - k_1(r')]\,\mathrm{d}r' = \Gamma[\varepsilon(1+\varepsilon^2)^{1/2} + \sinh^{-1}\varepsilon] \tag{6.56}$$

where

$$\Gamma = \frac{U_{12}^2}{\bar{k}|F_2 - F_1|} \tag{6.57}$$

Introducing this into (6.48) we have

$$\begin{aligned} M_{12} &= \frac{1}{2}\int_{-\infty}^{\infty} \frac{\mathrm{d}\varepsilon}{1+\varepsilon^2} \sin\{\delta^0 + \Gamma[\varepsilon(1+\varepsilon^2)^{1/2} + \sinh^{-1}\varepsilon]\} \\ &= \frac{\sin\delta^0}{2}\int_{-\infty}^{\infty} \frac{\mathrm{d}\varepsilon}{1+\varepsilon^2} \cos\{\Gamma[\varepsilon(1+\varepsilon^2)^{1/2} + \sinh^{-1}\varepsilon]\} \end{aligned} \tag{6.58}$$

where we have approximated the lower limit of the integral by $-\infty$. The integrand has poles at $\varepsilon = \pm\mathrm{i}$ and by contour integration with a cut along the imaginary axis starting at a pole one obtains the result

$$M_{12} = \left(\frac{\pi}{2} + J\right) \sin\delta^0 \exp\left(-\frac{\pi\Gamma}{2}\right) \tag{6.59}$$

where

$$J = \int_{1}^{\infty} \frac{\mathrm{d}\varepsilon}{\varepsilon^2 - 1} \sin[\cosh^{-1}\varepsilon - \varepsilon(\varepsilon^2 - 1)^{1/2}] \tag{6.60}$$

This integral depends on the value of Γ; it is zero when $\Gamma = 0$ and has the limit $-\pi/6$ at $\Gamma = \infty$. In fact this limit is reached rather rapidly and J may be taken as $-\pi/6$ for $\Gamma \geqslant 3$.

To obtain the transition probability from expression (3.100) we require the value of the zeroth-order Jost functions; these are equal to the asymptotic values of the amplitudes of the incoming distorted waves (A3.1). From (6.49),

$$B_{ii}^{(0)} = \tfrac{1}{2}k_i^{-1/2} \tag{6.61}$$

Hence, from (3.100) and (6.59) we obtain the transition probability

$$P_{12} = |S_{12}^2| = 4|M_{12}^2| = 4\left(\frac{\pi}{2} + J\right)^2 \sin^2\delta^0 \exp(-\pi\Gamma) \qquad 6.62$$

Although, as mentioned in Section 3.7, the **S** matrix in the distorted wave approximation is not unitary, when the adiabatic potentials are well separated (large Γ), P_{12} will be small and the error from non-unitarity will be unimportant. When Γ is large,

$$P_{12} \approx \frac{4\pi^2}{9} \sin^2\delta^0 \exp(-\pi\Gamma) \qquad 6.63$$

and taking the average value of $\sin^2\delta^0$,

$$P_{12} \approx \frac{2\pi^2}{9} \exp(-\pi\Gamma) \approx 2\exp(-\pi\Gamma) \qquad 6.64$$

Expression (6.62) is in general terms known as the Landau–Zener–Stueckelberg result although the precise form of the pre-exponential factor varies somewhat according to the method of derivation and the approximations used. The Landau–Zener formula itself refers to the probability of making a transition for a single passage across intersecting diabatic curves and is of the form

$$P_{LZ} = \exp(-\pi\Gamma) \qquad 6.65$$

In the Landau–Zener model we pass the crossing point at least twice in the atomic collision, on the way in and on the way out. We require a transition for one of these (probability P_{LZ}) but not the other (probability 1-P_{LZ}); hence the total Landau–Zener transition probability is

$$2P_{LZ}(1 - P_{LZ}) \qquad 6.66$$

which for a small transition probability is $2P_{LZ}$, which is consistent with (6.64).

There are three factors contributing to the magnitude of Γ. U_{12}, which is $2\mu V_{12}/\hbar^2$, determines the separation of the adiabatic potentials at the crossing point. When these potentials are well separated, the transition probability is very low. $|F_1 - F_2|$, which is $2\mu|V'_{11} - V'_{22}|/\hbar^2$, is proportional to the difference in slopes of the diabatic potentials at the crossing point. The transition probability will therefore be low if the diabatic potentials have similar slopes. Finally, k is proportional to the relative velocity of the atoms at the crossing point. If this is low, then again the transition probability will be low; i.e. for slow collisions there is a high chance of keeping to the adiabatic potentials.

Expression (6.62) and its approximations are invalid if Γ is very small; i.e. when U_{12} is small, k is large or $|F_2 - F_1|$ is large. In this case the dynamics are more appropriately confined to the diabatic curves and the correct solution will be obtained from the diabatic distorted wave approximation (6.31). The Landau–Zener transition probability for a double passage (6.66) does in fact go to zero as Γ approaches zero, but artificially, and one should not expect this formula to give a correct result in this limit. Nevertheless, when Γ is small, expression (6.66) is proportional to Γ and the diabatic limit can be shown to give the same result.[†]

There are many examples, particularly for molecular collisions, where there are multiple crossings of potential energy surfaces, so the question arises as to what happens when one has multiple crossings for atomic collisions. From the point of view of the exact solution of the coupled-channel equations there is no difficulty. However, it is always pleasant to have models that give simple analytical formulae. Nakamura[††] has in fact applied accurate formulae for the two-state problem to examples of three- and four-state crossings, and comparison with the exact coupled-channel results is very encouraging. His conclusion is that even when the individual crossings of diabatic curves are not well separated, the two-state formulae can be applied.

6.3 Classical trajectory methods

Collisions between molecules that promote electronic transitions will also produce changes in the vibrational and rotational energies of the molecules. Likewise, non-adiabatic reactive collisions will leave the products in a variety of rotational and vibrational states. Clearly, a full quantum mechanical treatment of such collisions will be very difficult—at least a greater level of difficulty than the corresponding processes that do not involve electronic transitions.

The classical trajectory treatments of such non-adiabatic processes also encounter difficulties, not in this case computational difficulties so much as conceptual difficulties. The problem is this: the electronic motion should be treated quantum mechanically and the nuclear motion classically. However, the nuclear motion depends on the electronic state, through the electronic potential energy surface, so the nuclear and electronic motions are essentially coupled. However, the coupling is not even-handed because in the Born–Oppenheimer approximation (6.8) the nuclear wave functions are specific to electronic eigenstates, not the other way round. Even if we derive an electronic wave function which represents the products of a collision emerging in a mixture of electronic states, there is no physical significance in the ro–vibrational motion on the averaged electronic potential appropriate to such mixed states. Only ro–vibrational states of electronically pure states are observable.

The inherent difficulty in mixing a quantum treatment of the electrons with a classical treatment of the nuclei led Miller and McCurdy[‡] to propose that the electrons should also be treated classically. The implications of this are, of course, rather dramatic—not least that the electronic energy spectrum is a continuum to which the semi-classical quantization rules would have to be applied to obtain discrete states. Although Miller and his coworkers have applied

[†]M. S. Child, *Molecular Collision Theory*, Academic Press, London and New York, 1974. Appendix D gives a thorough treatment of the various limiting cases that can be analysed in the two-state curve-crossing problem.

[††]H. Nakamura, *J. Chem.*, **87**, 4031 (1987).

[‡]W. H. Miller and C. W. McCurdy, *J. Chem. Phys.*, **69**, 5163 (1978).

these procedures to a number of different systems, they have not been widely adopted by others.

Classical trajectory methods for non-adiabatic processes divide themselves broadly into two sets, which we will call classical path and surface hopping. The former has its origin in an early model, developed by Mott for the collisions of atoms, in which the time-dependent Schrodinger equation for the electronic wave function is integrated along classical trajectories and the transition probabilities are calculated by averaging over impact parameters, b.† When the transition probability is low the potential governing the trajectory can be taken as the potential for the initial electronic state, and integrating (2.8) with this potential will give $r(t)$, which can then be used to deduce a time-dependent hamiltonian $H(r(t))$. The simplest version of the model assumes free-motion trajectories $[V(r)=0]$. If we take the zero of time at the distance of closest approach, then (see Figure 2.1)

$$r(t)=(b+r^2t^2)^{1/2} \qquad 6.67$$

where t runs from $-\infty$ to ∞. This version of the model is usually called the impact parameter method.

Coupling between the electronic and nuclear motion was first considered in detail by Nikitin and Pechukas and the first calculations were by Tully and Preston†† on the reaction of H^+ and D_2. Their work, as we shall see, led to the important surface hopping model.

During a trajectory the nuclear coordinates $\mathbf{Q}$ change with time and hence the electronic hamiltonian (6.3) can be considered to be time-dependent and its wave functions Φ satisfy the time-dependent Schrodinger equation

$$\mathrm{i}\hbar\frac{\partial\Phi(\mathbf{q},t)}{\partial t}=H_{\mathrm{e}}(q;Q(t))\Phi(\mathbf{q},t) \qquad 6.68$$

The function Φ can be expanded in the adiabatic wave functions (6.4) as follows:

$$\Phi(\mathbf{q},t)=\sum_j a_j(t)\Psi_j(\mathbf{q};\mathbf{Q}(t))\exp\left[-\frac{\mathrm{i}}{\hbar}\int V_j(\mathbf{Q}(t))\,\mathrm{d}t\right] \qquad 6.69$$

Initially $(t=0)$ the system will be in a single state $|a_i(0)|^2=1$ (say) and the populations of the states after the collision will be given by $|a_j(\infty)|^2$.

Substituting (6.69) into (6.68), multiplying by Ψ_i and integrating over $\mathbf{q}$ leads to the following set of coupled differential equations for the coefficients $a(t)$:

$$\begin{aligned}\dot{a}_i&=-\sum_j a_j\langle\Psi_i|\partial/\partial t|\Psi_j\rangle\exp\left[-\frac{\mathrm{i}}{\hbar}\int(V_j-V_i)\,\mathrm{d}t\right]\\&=-\sum_j a_j\langle\Psi_i|\partial/\partial\mathbf{Q}|\Psi_j\rangle\left(\frac{\partial\mathbf{Q}}{\partial t}\right)\exp\left[-\frac{\mathrm{i}}{\hbar}\int(V_j-V_i)\,\mathrm{d}t\right]\end{aligned} \qquad 6.70$$

†For a full account see D. R. Bates, *Atomic and Molecular Collision Processes*, Ch. 14, Academic Press, 1962.
††J. C. Tulley and R. K. Preston, *J. Chem. Phys.*, **55**, 562 (1971).

Equations (6.70) must be integrated numerically in parallel with Hamilton's equations for the nuclear motion (3.15). The remaining question to be answered is the appropriate potential energy function for this task. The average value of the electronic potential is, from (6.69),

$$\langle \Psi(\mathbf{q}, t)|H_e|\Phi(\mathbf{q}, t)\rangle = \sum_j |a_j|^2 V_j(\mathbf{Q}) \qquad 6.71$$

there being no terms in $a_j a_j'$, because we are using adiabatic functions which are eigenfunctions of H_e so that the potential energy matrix is diagonal; if we were using a diabatic expansion in (6.69) there would be cross terms as the diabatic potential is non-diagonal.

Figure 6.3 shows the results of classical path calculations for the excitation of Ne by He^+ using the same model potentials discussed earlier.† The pattern of oscillations in the partial wave expansion (Stueckelberg oscillations) in quite similar to that of the close-coupling results, but there is a cut-off at a slightly lower value of l and there are three fewer oscillations in all.

Tully and Preston applied the classical path model described above to the reaction of H^+ with D_2. This has an adiabatic reactive channel to give $D^+ + HD$ and non-adiabatic channels giving molecular ions, $H + D_2^+$ or $D + HD^+$. The two adiabatic potential surfaces show an avoided intersection when one of the three internuclear distances is approximately 1.3 Å. This is due to the fact that the diatomic potentials for H_2 and H_2^+, when plotted with the same dissociation limit, have a crossing point at this distance. There are 'seams' of such avoided crossings on the surface, the separation of the two surfaces depending on the values of the other two internuclear distances.

The most important conclusion from Tully and Preston's calculations was that the electronic wave functions only showed significant change when the trajectories crossed seams, and that motion perpendicular to a seam was much more effective than motion parallel to a seam. This is due to the fact that the coupling terms $\langle \Psi_i|\partial/\partial \mathbf{Q}|\Psi_j\rangle$ which appear in (6.70) have their largest values when $\mathbf{Q}$ is perpendicular to a seam. Their other conclusion was that the Landau–Zener formula (6.65) when applied to the crossing point gave a good approximation to the transition probabilities.

We have already noted that the weakness of the classical path method is that the emerging trajectories are on potentials which are electron-state averages (6.71), and the ro–vibrational motion on such potentials has no connection with the observable ro–vibrational distributions of the pure states. Tully and Preston, as a result of their classical path calculations, were therefore led to propose a

†J. P. Braga, L. J. Dunne and J. N. Murrell, *Chem. Phys. Lett.*, **120**, 147 (1985). The electronic populations in these calculations were deduced from the elements of the density matrix $\rho_{ij} = a_i^* a_j$ whose time variation is given by the quantum mechanical Liouville equation

$$i\hbar \frac{\partial \boldsymbol{\rho}}{\partial t} = [\mathbf{V}, \boldsymbol{\rho}]$$

For a two-state problem it is probably easier to use this rather than (6.70). The average value of the potential in this formulation is Trace $(\mathbf{V}, \boldsymbol{\rho})$.

model in which trajectories were calculated wholly on the adiabatic surfaces, but when a trajectory reached a seam it either continued on the same surface or it was transferred to the neighbouring adiabatic surface where it continued to evolve until another seam was reached. The probability of making the 'surface hop' was to be calculated by applying the Landau–Zener formula (6.65).

Tully and Preston proposed two methods for computing the product-state distribution from a set of trajectories. The first, which they called the 'ants' procedure, was to bifurcate a trajectory each time it reached a seam. If there is a probability P for transfer, then at the first crossing a trajectory is continued on the new surface and given a weighting P, but the original trajectory is also continued and given a weighting $(1 - P)$. These trajectories will have further bifurcations each time they reach a crossing point; the column of ants splits into two, with appropriate numbers, at each crossing point. The final trajectories emerge on one or other of the adiabatic surfaces with weighting factors that are multiples of P_i or $(1 - P_i)$ accumulated at each crossing point. These trajectories can be analysed for their rotational and vibrational energies in the usual way.

The above procedure becomes impractical if trajectories pass through seams many times, because a large number of very low probability trajectories have eventually to be computed; 'ants' are swarming all over the surface. Under these circumstances the 'anteater' procedure is recommended. When the trajectory reaches the first seam the Landau–Zener probability P_1 is calculated and this is compared with a computer-generated random number ζ_1 between 0 and 1. If $P_1 > \zeta_1$ the trajectory jumps to the new surface and evolves on this; if $P_1 < \zeta_1$ the trajectory remains on the original surface. This procedure is repeated at each crossing point and on a single trajectory emerges on one or other of the adiabatic surfaces with weight unity.

When a trajectory jumps to a new surface its potential energy changes discontinuously and without further correction the total energy would not be conserved. To correct for this, Tully and Preston proposed that the component of the momentum perpendicular to the seam should be changed. The recipe they gave for this was not, however, completely general and the problem was later examined in detail by Miller and George.† They analysed trajectories on a stepped surface

$$V(\mathbf{Q}) = V_1(\mathbf{Q})h[f(\mathbf{Q})] + V_2(\mathbf{Q})h[-f(\mathbf{Q})] \qquad 6.72$$

where $h(x)$ is the step function:

$$h(x) = \begin{cases} 1, & x > 0 \\ 0, & x < 0 \end{cases} \qquad 6.73$$

and where $f(\mathbf{Q}) = 0$ defines the seam. As the motion is on a time-independent (albeit non-analytic) potential, the total energy and angular momenta are conserved. By integrating Hamilton's equations over infinitesimal time steps on

†W. H. Miller and T. F. George, *J. Chem. Phys.*, **56**, 5637 (1972).

either side of the seam, Miller and George were able to show that the initial change in the momentum was indeed perpendicular to the seam, as proposed by Tully and Preston. For the A + BC collision, the following expressions have been given for the changes in the six momenta components in a centre-of-mass coordinate system following transfer from potential V_1 to V_2:†

$$\Delta P_i = -\left(\frac{\partial f}{\partial Q_i}\right)\frac{A}{B}\left\{1 - \left[1 + \frac{2(V_1 - V_2)B}{A^2}\right]^{1/2}\right\} \tag{6.74}$$

where

$$A = \sum_{i=1,3}\frac{P_i}{\mu_{BC}}\left(\frac{\partial f}{\partial Q_i}\right) + \sum_{i=4,6}\frac{P_i}{\mu_{A,BC}}\left(\frac{\partial f}{\partial Q_i}\right) \tag{6.75}$$

$$B = \sum_{i=1,3}\frac{1}{\mu_{BC}}\left(\frac{\partial f}{\partial Q_i}\right)^2 + \sum_{i=4,6}\frac{1}{\mu_{A,BC}}\left(\frac{\partial f}{\partial Q_i}\right)^2 \tag{6.76}$$

all qualities being evaluated at the seam.

The surface hopping procedure can also be carried out on diabatic surfaces which have real, rather than avoided, intersections. In such cases there is no need to make changes in the momenta, as is clear by making $V_2 = V_1$ in (6.74).

The surface hopping calculations by Tully and Preston on the H_3^+ surface were surprisingly successful in view of the approximate nature of their potential energy functions and the crudity of the model. Figure 6.5 shows a comparison between the total cross-sections calculated at three different collision energies and experimental results by Ochs and Teloy; earlier experimental values were in poorer agreement with the calculations. The calculations also gave a good account of the isotope effect on the dynamics, which show that in the formation of molecular ions simple charge transfer has a larger cross-section than reactive charge transfer.

There are several situations in which the Tully–Preston procedure will prove unreliable. For example, if the transition probability at a seam is very high or low, then the Landau–Zener formula is likely to be inaccurate. Truhlar and Muckerman suggest that in this case one should calculate the transition probability by locally using the classical path method described earlier. This procedure can also be used when the surfaces show no clearly defined avoided intersections. In this case the points at which surface hopping should occur can be taken as those at which the classical path integration reaches a certain specified probability.††

However, the most serious criticism that one can make of the surface hopping procedure is that it accumulates probabilities rather than probability amplitudes. Thus, although the electronic transition probabilities at a seam are being calculated by quantum or semi-classical means, they are being accumulated by a simple classical model. There is, for example, no interference in the transition probabilities from multiple crossings of the same or different seams. There are,

†D. G. Truhlar and J. T. Muckerman; in R. B. Bernstein (ed.), *Atom–Molecule Collision Theory. A Guide for the Experimentalist*, Ch. 16, Plenum Press, New York, 1979.
††N. C. Blais and D. G. Truhlar, *J. Chem. Phys.*, **79**, 1334 (1983).

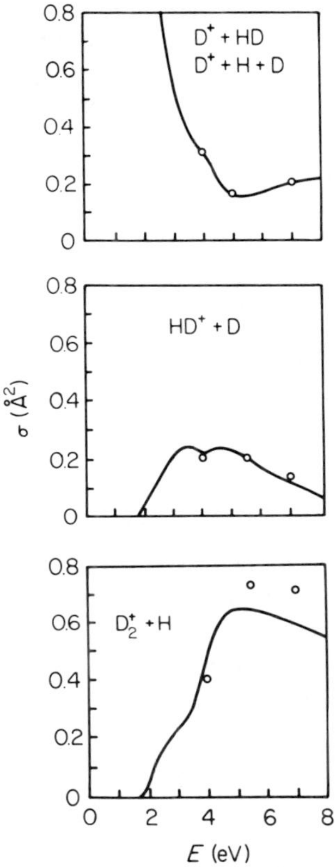

Figure 6.5 Cross-sections for the collision of $H^+ + D_2$ into various reaction channels. The circles show surface hopping results (*J. R. Krenos, R. K. Preston, R. Wolfgang and J. C. Tully*, J. Chem. Phys., **60**, *634*, *1974*) and the full curves show experimental results (*G. Ochs and E. Teloy*, J. Chem. Phys., **61**, 4930, 1974)

in effect, no Stueckelberg-type oscillations from the surface hopping model.

The question is how does one maintain the best feature of the surface hopping procedure, namely motion on adiabatic surfaces, but dispense with the weaknesses (hopping at discrete points, discontinuous momenta and accumulation of probability rather than amplitude)? The most elegant attempt is undoubtedly the proposal of Miller and George to extend the semi-classical path integral approach, which we described in Section 3.6 in regard to inelastic scattering, to non-adiabatic processes. We remind the reader of the essence of this method. One seeks classical trajectories that go from specified states (described by sets of action quantum numbers) of the reactants to specified states of the products, and a probability amplitude is calculated for such trajectories from the classical action integral and the use of expressions such as (3.80). The total transition amplitude is obtained by summing the contributions from each trajectory leading from specified initial to specified final states.

In extending this method to non-adiabatic processes, we must effectively add one more quantum number to the description of the initial and final states, and that is a quantum number labelling the electronic states, or the adiabatic surfaces associated with these states.

The problem is how to find trajectories that pass smoothly from one surface to another, for this can only be done if the two surfaces cross. If we have a diabatic representation of two surfaces, then from (6.19), the adiabatic surfaces will only cross if

$$(V'_{11} - V'_{22})^2 + 4(V'_{12})^2 = 0 \qquad 6.77$$

As the function is a sum of two squared terms, we would normally consider that this condition is satisfied only if both terms are zero. However, a mathematical solution to (6.77) is

$$V'_{11} - V'_{22} = \pm 2\mathrm{i}V'_{12} \qquad 6.78$$

so that surfaces which only show avoided crossing in real space can cross in complex space.

One can only get complex values of the potential by having complex coordinates, and this implies that their conjugate momenta are also complex. Thus the extension of nuclear dynamics into complex space doubles the number of independent nuclear variables because the real and imaginary components of the coordinates and momenta must be treated as independent variables. Moreover, if we start a trajectory in real space it can only move into complex space if the time variable is also allowed to be complex.

The method of computing trajectories in complex space had earlier been proposed by Miller[†] to handle inelastic and reactive collisions on a single surface where certain channels could only be opened by allowing trajectories to tunnel through potential barriers. Within a barrier the kinetic energy is negative so complex values of the momenta are required. Integration of Hamilton's equations in the complex region leads to problems of divergencies (angle variables in complex space lead to exponential behaviour) and this makes the method difficult to use. However, the main problem with the method is the one we mentioned in Section 3.6 for inelastic scattering, which is that when there is more than one internal coordinate it is difficult to find trajectories that have specified initial and final quantum numbers. This elegant classical path solution to the problem of non-adiabatic processes has therefore not been greatly used.

[†]W. H. Miller, *Chem. Phys. Lett.*, **2**, 431 (1970). W. H. Miller and T. F. George, *J. Chem. Phys.*, **56**, 5668 (1972).

Chapter 7

Scattering from surfaces

7.1 Introduction

The scattering of atoms or molecules from surfaces can exhibit all the phenomena associated with scattering by other molecules. There can, for example, be a change in the internal energy or there may be breaking of a chemical bond. There are, in addition, two other phenomena. The first is the appearance of a diffraction pattern in the scattered beam if the surface has a regular periodic array of atoms. This arises from the interference of waves scattered by different but equivalent atoms in the surface. The second is that the atom or molecule may stick to the surface by giving up its translational energy to the vibrations of the lattice.

There are two motives for the study of molecule–surface scattering. The first is to probe the structure of the surface, or more precisely the nature of the potential acting between the atom or molecule and the surface. The scattering of, say, monoenergetic He atoms from a surface is a probe in the same sense that X-rays, neutrons or electrons scattered from a surface are probes. The differences between these experiments arise from different potentials and particularly different penetration depths of the beam into the crystal. Low energy beams of atoms will normally only probe the first surface layer of atoms.

The second motive is to understand the specific interaction between the atom or molecule and the surface because this is relevant to such important physicochemical phenomena as catalysis, ion implantation and crystal growth.

If of course the surface is irregular, not having even short-range order, then there will be no diffraction pattern and at best the scattering would be amenable to some statistical analysis. In the discussion that follows we consider only idealized surfaces with two-dimensional regular periodicity. There is one further considerable simplification that can be made and that is to take the atoms in the surface to be fixed at their lattice sites. This limits the phenomena that can occur in one important respect because colliding atoms or molecules cannot be trapped on the surface as there is no mechanism for transferring their translational energy to lattice phonons. Molecules, of course, may become temporarily trapped if most of their translational energy is converted to rotational or vibrational energy in the collision; this is a resonance effect analogous to that which can occur in molecule–molecule collisions, but permanent trapping by a 'frozen' surface is not possible.

The theory of atom or molecule scattering with surfaces that have moving atoms is very difficult, but of course very interesting. The fixed-surface model is probably adequate if one is just interested in surface structure, but it will have little relevance to problems such as catalysis or crystal growth.

We deal first with the relatively simple problem of the scattering of atoms from a fixed periodic lattice. This will be followed by an examination of molecule scattering from fixed lattices and finally we look briefly at the problem of scattering from a vibrating surface.

7.2 Atom–rigid surface scattering

We adopt the convention that the surface is in the (x, y) plane and define the two-dimensional periodicity of the lattice by the translational symmetry of the atom–surface potential

$$V(x, y, z) = V(x + ma_x, y + na_y, z) \tag{7.1}$$

where a_x and a_y are the lattice constants and m and n are positive or negative integers. Such a periodic function can be written as a Fourier expansion:

$$V(x, y, z) = \sum_{m,n=-\infty}^{\infty} V_{mn}(z) \exp(img_x x + ing_y y) \tag{7.2}$$

where $\mathbf{g}$ is called the reciprocal lattice vector; its components are given by $g_x = 2\pi/a_x$, $g_y = 2\pi/a_y$. We can write this in a briefer notation

$$V(x, y, z) = \sum_{m,n=-\infty}^{\infty} V_{mn}(z) \exp(\mathrm{i}\mathbf{G}_{mn} \cdot \mathbf{R}) \tag{7.3}$$

where $\mathbf{G}_{mn}$ has components mg_x and ng_y, and $\mathbf{R}$ is the (x, y) vector.

The Schrodinger equation for atom–surface collisions is

$$\left[-\frac{\hbar^2}{2\mu} \nabla^2 + V(x, y, z) \right] \Psi(x, y, z) = E\Psi(x, y, z) \tag{7.4}$$

It will be recalled that the Schrodinger equation for atom–atom or atom–molecule scattering can be solved by making an expansion in a set of known functions (the basis), taking matrix elements in this basis, and finally integrating equations in just the collision or reaction coordinate r. This leads to the partial wave expansion for elastic scattering or the coupled-channel equations for inelastic scattering. For atom–surface scattering we employ the same technique.

As will be clear shortly, given the form of the expansion used for the potential (7.3), the most appropriate form for the wave function is that which is called the Laue representation:

$$\Psi(x, y, z) = A^{-1/2} \sum_{m,n=-\infty}^{\infty} \phi_{mn}(z) \exp[\mathrm{i}(\mathbf{K} + \mathbf{G}_{mn}) \cdot \mathbf{R}] \tag{7.5}$$

where $\mathbf{K}$ is the projection of the wave vector $\mathbf{k}$ ($k^2 = 2mE/\hbar^2$) of the atom beam in the (x, y) plane, A is the area of the unit cell and the factor $A^{-1/2}$ has been introduced so that the basis functions are normalized for integration over one unit cell.

Introducing (7.3) and (7.5) into (7.4) and multiplying by the complex conjugate of one of the basis functions of the expansion gives the equation

$$A^{-1}\mathrm{e}^{-\mathrm{i}(\mathbf{K}+\mathbf{G}_{pq})\cdot\mathbf{R}}\left\{-\frac{\mathrm{d}^2}{\mathrm{d}x^2}-\frac{\mathrm{d}^2}{\mathrm{d}y^2}-\frac{\mathrm{d}^2}{\mathrm{d}z^2}+\frac{2\mu}{\hbar^2}\sum_{mn}V_{mn}(z)\mathrm{e}^{\mathrm{i}\mathbf{G}_{mn}\cdot\mathbf{R}}-k^2\right\}$$

$$\times\sum_{mn}\phi_{mn}(z)\mathrm{e}^{\mathrm{i}(\mathbf{K}+\mathbf{G}_{mn})\cdot\mathbf{R}}=0 \qquad 7.6$$

We now integrate (7.6) over the x and y coordinates. Because the Fourier exponential functions are an orthonormal set the terms which are independent of the potential give only a diagonal element ($mn = pq$), but the potential gives rise to off-diagonal or coupling terms. We therefore have a set of coupled-channel equations of the form

$$\left\{-\frac{\mathrm{d}^2}{\mathrm{d}z^2}-[k^2-(\mathbf{K}+\mathbf{G}_{mn})^2]+\frac{2\mu}{\hbar^2}V_{mn,mn}(z)\right\}\phi_{mn}(z)$$

$$+\frac{2\mu}{\hbar^2}\sum_{pq\neq mn}V_{pq,mn}(z)\phi_{pq}(z)=0 \qquad 7.7$$

the factor A^{-1} having disappeared as a result of our normalization condition. The matrix elements of $V(z)$ are given by

$$V_{pq,mn}(z)=A^{-1}\int_{-\infty}^{\infty}\mathrm{e}^{-\mathrm{i}(\mathbf{K}+\mathbf{G}_{pq})\cdot\mathbf{R}}\sum_{m'n'}V_{m'n'}(z)\mathrm{e}^{\mathrm{i}\mathbf{G}_{m'n'}\cdot\mathbf{R}}\mathrm{e}^{\mathrm{i}(\mathbf{K}+\mathbf{G}_{mn})\cdot\mathbf{R}}\,\mathrm{d}x\,\mathrm{d}y$$

$$=A^{-1}\sum_{m'n'}V_{m'n'}(z)\int_{-\infty}^{\infty}\mathrm{e}^{\mathrm{i}(\mathbf{G}_{mn}+\mathbf{G}_{m'n'}-\mathbf{G}_{pq})\cdot\mathbf{R}}\,\mathrm{d}x\,\mathrm{d}y \qquad 7.8$$

The individual integrals occurring in this summation will be zero unless $m + m' - p = 0$ and $n + n' - q = 0$; in other words, the channels mn and pq are directly coupled by just one term in the expansion of the potential (7.3) which satisfies these two conditions. We can therefore write

$$V_{pq,mn}(z)=\sum_{m'n'}V_{m'n'}(z)\delta(m+m'-p)\delta(n+n'-q) \qquad 7.9$$

A commonly used potential for atom–surface scattering is that proposed by Lennard-Jones and Devonshire:

$$V(x,y,z)=V_0(z)+V_1(z)\left[\cos\left(\frac{2\pi x}{a_x}\right)+\cos\left(\frac{2\pi y}{a_y}\right)\right] \qquad 7.10$$

which, when written as (7.3), is

$$V(x,y,z)=V_0(z)+\tfrac{1}{2}V_1(z)(\mathrm{e}^{\mathrm{i}\mathbf{G}_{10}\cdot\mathbf{R}}+\mathrm{e}^{\mathrm{i}\mathbf{G}_{01}\cdot\mathbf{R}}+\mathrm{e}^{\mathrm{i}\mathbf{G}_{-10}\cdot\mathbf{R}}+\mathrm{e}^{\mathrm{i}\mathbf{G}_{0-1}\cdot\mathbf{R}}] \qquad 7.11$$

The coupled-channel equations (7.7) in this case will be

$$\left\{-\frac{d^2}{dz^2}-[k^2-(\mathbf{K}+\mathbf{G}_{mn})^2]+\frac{2\mu}{\hbar^2}V_0(z)\right\}\phi_{mn}(z)$$

$$+\frac{\mu}{\hbar^2}V_1(z)(\phi_{mn+1}+\phi_{mn-1}+\phi_{m+1n}+\phi_{m-1n})=0 \qquad 7.12$$

so that each channel (m, n) will be coupled equally to four other channels.

To obtain **S**-matrix elements we must apply boundary conditions to the solutions of the coupled-channel equations (7.8). These are similar to those used for the collinear collision of an atom and a diatomic molecule in Section 3.5 [cf. (3.72)]. If there is no penetration into the solid [$V(z)\to\infty$ for $z\to-\infty$] the channel wave functions will all approach zero as $z\to-\infty$. As the potential approaches zero in the separate atom–surface limit ($z\to\infty$) we have, for scattering from the incident diffraction channel $(m'n')$ to the reflected diffraction channel (mn),

$$\phi_{mn}\sim e^{-id_{m'n'}z}\delta_{mm'}\delta_{nn'}-\left(\frac{d_{m'n'}}{d_{mn}}\right)^{1/2}e^{id_{mn}z}S_{mn,m'n'} \qquad 7.13$$

where

$$d_{mn}^2=\mathbf{k}^2-(\mathbf{K}+\mathbf{G}_{mn})^2 \qquad 7.14$$

The channel wave functions have been normalized so that $|\mathbf{S}|^2$ gives the scattering probabilities.

Now $\mu\hbar\mathbf{K}$ is the momentum of the incident atom beam in the plane of the surface and $\mu\hbar\mathbf{G}_{mn}$ can be considered as the momentum acquired in this plane as a result of the surface potential. By definition the incident channel is that for which $m'=n'=0$, and this is called the specular channel. The probabilities of scattering into the diffraction channel $(m\,n)$ (the diffraction intensities) are

$$P(mn)=|S_{mn,00}|^2 \qquad 7.15$$

$P(00)$ is the intensity of the 'specular' diffraction peak.

If the incident beam comes in at the spherical polar angles (θ_i, ϕ_i) its **K** vector has components

$$K_x=k\sin\theta_i\cos\phi_i$$
$$K_y=k\sin\theta_i\sin\phi_i \qquad 7.16$$

The (m, n) diffraction peak then appears at angles $(\theta_f^{mn}, \phi_f^{mn})$ given by

$$K_x^m=K_x+mg_x=k\sin\theta_f^{mn}\cos\phi_f^{mn}$$
$$K_y^n=K_y+ng_y=k\sin\theta_f^{mn}\sin\phi_f^{mn} \qquad 7.17$$

with relative intensities $P(mn)$.

Scattering of an atom from a rigid surface is an elastic event because the total energy of the atom beam is conserved. However, as in atom–atom or central field scattering, the momentum of the beam can change direction. Of course, $(\mathbf{K}+\mathbf{G}_{mn})$

cannot be greater than $\mathbf{k}$, and channels $(m\,n)$ which violate this condition will be closed channels in the coupled-channel equations (7.7). There is therefore a finite number of diffraction peaks (open channels) available for the scattered beam.

If a surface possesses symmetry this can be used to reduce the number of coupled-channel equations. As the specular channel (0, 0) is totally symmetric it is only necessary to consider coupling to other totally symmetric channels. For example, for a square lattice (C_{4v} symmetry) with p^2 diffraction channels there are roughly $p^2/8$ combinations of A_1 type such as

$$A_1(\alpha, \beta) = [(\alpha, \beta) + (\alpha, -\beta) + (-\alpha, \beta) + (-\alpha, -\beta) + (\beta, \alpha) + (\beta, -\alpha) + (-\beta, \alpha) + (-\beta, -\alpha)]/8^{1/2} \qquad 7.18$$

and the computing time required to solve the equations may be reduced by a factor of about one hundred by making use of this symmetry.

It is also possible to use a sudden approximation on these equations. If the incident wave vector $\mathbf{k}_z$ is large compared with $|\mathbf{G}_{mn}|$ for all channels that significantly contribute to scattering, then one can replace d_{mn}^2 in (7.14) by an average value d^2 and, in some cases, find a matrix transformation (independent of z) which diagonalizes the potential energy matrix appearing in (7.7). The $\mathbf{S}$-matrix in this approximation is then given by a transformation analogous to (3.111).

To perform classical dynamical calculations on atom–surface scattering one must run batches of trajectories for fixed incident angles and energies with varying initial positions with respect to the centre of the unit cell. In the directions parallel to the surface the position must span the range $(\pm\frac{1}{2}a_x, \pm\frac{1}{2}a_y)$; the initial vertical distance from the surface is arbitrary, but must be sufficiently large that the potential at this distance can be taken as zero. Varying the initial position is equivalent to varying the impact parameter for atom–atom scattering.

There may be more than one trajectory scattering into a particular angle, as is the case for atom–atom scattering, but as there is no interference between trajectories, there will be no oscillatory diffraction pattern. However, there will frequently be classical rainbows and unlike the atom–atom case, these are not necessarily due to attractive regions of the potential. For example, Figure 7.1 shows classical scattering from a one-dimensional hard corrugated potential. The rainbows for such a potential arise from incident trajectories which impact at points on the surface where the slope of the corrugation is changing most slowly, i.e. from the points of inflection of the corrugation.

Table 7.1 compares calculated intensities of the diffraction peaks for normal incident beams ($k_x = k_y = 0$) which have been obtained by three methods: exact coupled channel, sudden approximation and classical. The classical results were obtained by assigning continuous variables (m, n) to the trajectories of the reflected beam using (7.17) and integerizing these in the usual way.

Two examples are given: one, Ne on a W(110) surface represents a relatively heavy atom on a weakly corrugated surface, and the other, He on LiF(001), is a light atom on a strongly corrugated surface. In both cases the atom–surface

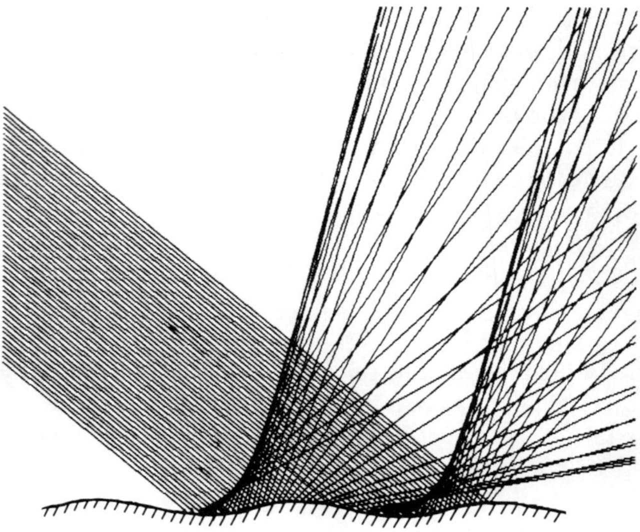

Figure 7.1 Classical scattering from a one-dimensional hard corrugated surface. The rainbows arise from impacts at the points of inflection of the surface

Table 7.1 Comparison of classical, sudden and exact coupled-channel calculations for the diffraction intensities of two systems. The channels, $(\pm m, \pm n)$ and $(\pm n, \pm m)$, have equal intensities. Dashes indicate closed channels. (*Data from A. T. Yinnon, S. Bosanac, R. B. Gerber and J. N. Murrell*, Chem. Phys. Lett., **58**, *364, 1978.*)

	He/LiF			Ne/W		
mn	Classical	Sudden	Exact	Classical	Sudden	Exact
00	0.0172	0.0059	0.0208	0.1380	0.4002	0.4016
10	0.0151	0.0041	0.0098	0.1168	0.1096	0.1099
20	0.0205	0.0182	0.0370	0.0	0.0064	0.0061
30	0.0170	0.0143	0.0186	0.0	0.0002	0.0001
40	0.0	0.0095	0.0008	—	—	—
11	0.0160	0.0244	0.0196	0.0987	0.0301	0.0299
22	0.0382	0.0309	0.0705	0.0	0.0001	0.0001
33	0.0160	0.0352	0.0014	—	—	—
21	0.0230	0.0095	0.0091	0.0	0.0018	0.0017
31	0.0176	0.0012	0.0151	0.0	0.0001	0.0
32	0.0208	0.0236	0.0184	—	—	—
41	0.0	0.0044	0.0008	—	—	—
42	0.0	0.0173	0.0002	—	—	—

potential has the form (7.10) with $V_0(z)$ being a Morse potential and $V_1(z)$ an exponential repulsion. The beam energies are approximately the same, 0.06 eV, but because of the greater mass of the Ne there are many more open channels for Ne/W than for He/LiF, 365 and 69 respectively.

Because of the weak corrugation for Ne/W there are fewer classically allowed diffraction peaks. These peaks can nevertheless appear weakly in the quantum calculations. This is analagous to the appearance of classically forbidden transitions for collinear atom–diatomic molecule vibrational excitation shown in Table 3.1, and illustrates the effect of tunnelling.

The sudden approximation appears to be quite good for the Ne/W system. The condition $k_z \gg |G_{mn}|$ holds for the classically allowed channels; this is because the corrugation is small and in consequence the maximum change in $|\mathbf{G}|$ is small. Surprisingly the classical result is better for He/LiF. This is probably because there are more classically allowed open channels and the 'boxing' procedure will be more reliable for dense diffraction peaks. For completion we note that a semi-classical analysis can also be made on the classical trajectories in exactly the same manner as it was applied to vibrationally inelastic atom–diatomic molecule collisions. The channel integers (m, n) obtained from classical trajectories are functions of the initial coordinates, as explained earlier, so that a many-dimensional analogue of Figure 3.4 could be constructed and the specific trajectories arriving at integer (m, n) could be combined with an appropriate phase, as in expression (3.82).

For potentials with high corrugation, or low incident angle beams, there may be inaccessible points on the potential, and there may also be multiple collisions. It is also possible for sufficient momentum to be removed from the z direction for the atom to be temporarily trapped within an attractive atom–surface potential. This is known as selective adsorption and is a Feshbach-type resonance. Classically these temporarily trapped trajectories would have substantial momenta in the (x, y) plane, which would remain until later interactions return momentum to the z direction. Typical lifetimes of these resonance states are 10^{-12} s with the atom travelling about 10 Å on the surface before being ejected. The atom cannot become totally trapped unless energy can be given up to the lattice vibrations.

However, the quantum mechanical picture is more illuminating because the resonances are linked to the discrete vibrational levels of the atom attached to the surface. When the energy of a bound state corresponds to that of a closed diffraction channel there can be resonances in the open channels. Resonances occur when there is a strong influence of closed channels on the solution of the coupled-channel equations. For a weakly corrugated surface the condition for a resonance is [cf. (7.7)],

$$k^2 - (K + G_{mn})^2 = \frac{2\mu}{\hbar^2}\varepsilon_p \qquad 7.19$$

where ε_p is a bound state of the $V_{00}(z)$ potential.

Figure 7.2 shows the selective adsorption resonances for He scattered from a

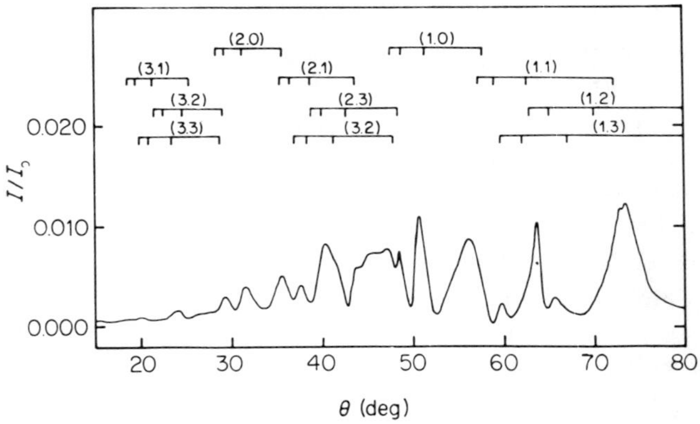

Figure 7.2 Specular reflection of He scattered from a Xe monolayer on graphite and the positions of calculated selective adsorption resonances

Xe monolayer on graphite.[†] The Xe in this case is sitting above the centre of each graphite hexagon so that these resonances measure rather directly the He–Xe potential. The figure shows the effect of varying the incident angle θ (ϕ kept constant at zero) for constant incident energy on the intensity of the specular peak [channel (0, 0); an alternative experiment would be to vary the energy for the fixed incident angle]. Also shown in the figure are the calculated resonances for each diffraction channel based on an atom–surface potential such as (7.2).[††] This was constructed by taking a He–Xe pair potential and a long-range He–graphite attractive potential proportional to R^{-3}. The Fourier expansion of this contained the terms V_{00}, V_{10}, V_{20}, V_{11} and V_{21}.

There are four resonances for each closed diffraction channel and, from left to right in the figure, these correspond to the threshold, and vibrational levels of $V_{00}(z)$ at -0.67, -2.02, and -5.02 meV. These resonances are rather dense and it is difficult to see a direct connection between a particular resonance and the peaks in the scattering intensity.

7.3 More complicated problems

In recent years there has been a large increase in experimental studies of molecule–surface scattering because of the sensitivity of laser-induced fluorescence for identifying the ro–vibrational states of molecules, and the increased resolution of time-of-flight analysis of the reflected beams. Clearly the theoretical analysis is very demanding because in the quantum mechanical coupled-channel method we must identify channels both by their diffraction integers and by their ro–vibrational quantum numbers. The problem therefore has the combined complexity of atom–molecule and atom–surface collisions, and 'sudden' approximations to the coupled-channel equations are almost essential for their solution. The only saving is that the **S**-matrix elements give probabilities directly,

[†]G. Bracco, MSc Thesis, University of Geneva (1981).
[††]J. M. Hutson and C. Schwarz, *J. Chem. Phys.*, **79**, 5179 (1983).

unlike atom–molecule rotational inelasticity where a partial wave expansion is required to give cross-sections.

The classical trajectory method, however, has no greater computational difficulty than for atom–molecule collisions and is relatively easy to apply. Although one can question its reliability, this should be no worse orse than for atom–molecule scattering. The method will become more reliable the more averaging there is in the incident beam conditions or in the reflected beam state resolution and angular resolution.

Molecule–surface scattering will be strongly influenced by rotational rainbows, which arise by a mechanism which is entirely analogous to that for atom–molecule scattering. They arise from zeros of the jacobian (4.19) of final quantum numbers versus initial phases. It is always tempting to determine empirical potentials for the scattering process by fitting calculations to experimental rainbows. Although this has been done for several systems†, it is not yet clear that the computational methods are sufficiently accurate to produce reliable potentials by this method.

If there is a large transfer of translational energy to rotational or vibrational energy on impact with the surface then a molecule can become temporarily trapped even by a rigid surface. This is analogous to the occurrence of rotational or vibrational resonances in atom–molecule collisions. In the context of molecule–surface collisions one uses the terms rotationally or vibrationally mediated selective absorption.

By analogy to (7.19) the condition for a rotational resonance is

$$k^2 - (K + G_{mn})^2 - \frac{2\mu}{\hbar^2}\,\Delta E_{\text{rot}} = \frac{2\mu}{\hbar^2}\,\varepsilon_{\text{p}} \qquad 7.20$$

where ΔE_{rot} is the change in rotational energy between the incident channel and the closed channel connected to the resonance. These resonances are normally difficult to observe because of the high density of rotational levels; H_2 is clearly the most favourable case and resonances have been seen for scattering from Ag(111) surfaces.††

Transfer of energy between an atom or molecule and a surface, trapping on the surface being one possible outcome, depends on the vibrations of the atoms of the surface.

The modes of vibration of a lattice are called phonons and, in contrast to the discrete vibrational levels of a molecule, have an energy continuum. Some phonons have their maximum amplitude at the surface and can be called surface waves, and it is these that can be strongly excited by atomic or molecular collisions.

It is generally agreed that most cases of energy exchange to a surface is a multiphonon process, the exception being for light atom collisions. Weare‡ has

†J. C. Tully, in D. Langreth and H. Suhl (eds.), *Many-Body Phenomena at Surfaces*, Academic Press, 1984.
††C. F. Yu, K .B. Whaley, C. S. Hogg, and S. J. Sibener, *Phys. Rev. Lett.*, **51**, 2210 (1983).
‡J. H. Weare, *J. Chem. Phys.*, **61**, 2900 (1974).

shown that the ratio of two-phonon to one-phonon transitions is given by the factor

$$\beta = \left(\frac{2\mu^2 k_z^2}{M\hbar^2}\right)\left(\frac{T}{k_B\theta_D^2}\right) \qquad 7.21$$

where μ is the mass of the colliding atom and M the mass of a surface atom. T is the surface temperature (assumed high compared with $k_B\theta_D$) and θ_D the Debye temperature of the solid that measures the width of the phonon band; k_B is Boltzmann's constant. In typical experimental conditions for He beams $\beta \approx 0.05$.

If we take a classical view of the collision we can picture, in the simplest case, an incoming atom colliding with a single atom of the surface. The transfer of energy will therefore be local whereas the phonons are delocalized over the lattice. The energy will be transferred to several modes of vibration in such a collision. Single phonon excitation is analogous to exciting by one quantum a normal mode of a molecule; multiphonon processes are analogous to exciting overtones or combination bands. At higher temperatures the lattice will be in energy states in the middle of a phonon band and the probability of multiphonon jumps will be increased.

The low energy scattering of He from a surface appears to be predominantly a single phonon process, and by measuring the loss of energy in the reflected beam, using time-of-flight methods, one can determine the phonon distribution curve. For energy balance

$$\frac{\hbar^2(\mathbf{k}')^2}{2\mu} = \frac{\hbar^2\mathbf{k}^2}{2\mu} \pm \hbar\omega(\mathbf{Q}) \qquad 7.22$$

where $\mathbf{k}$ is the initial and $\mathbf{k}'$ the final wave vectors, and $\pm\hbar\omega(\mathbf{Q})$ the energy change for addition or removal of a phonon of wave vector $\mathbf{Q}$. For momentum conservation in the surface plane

$$\mathbf{K}' = \mathbf{K} + \mathbf{G} + \mathbf{Q} \qquad 7.23$$

where $\mathbf{K}$ and $\mathbf{K}'$ are the wave vectors of the atomic beam in the surface plane and $\mathbf{G}$ the reciprocal lattice vector appropriate to the particular reflection angle being measured. From experiment† one can determine $\omega(\mathbf{Q})$, which is called the phonon dispersion curve.

To obtain a detailed interpretation of the inelastic scattering of an atom by a surface with one-phonon excitation, one must first be able to solve the rigid-surface problem. We have already seen in Section 7.2 that this is not a difficult task using the coupled-channel method. The resulting scattering wave functions can then be used as the zeroth-order functions in the distorted wave Born approximation (3.100). The perturbation terms, $U_{ij}(X)$ in (3.100), are in this case matrix elements between the phonon states of the interaction potential between

†For example, J. P. Toennies, in G. Benedek and U. Valbusa (eds.), *Dynamics of Gas–Surface Interactions*, Springer-Verlag, 1982.

the incoming atom and the non-rigid surface; this potential depends on the coordinates of the surface atoms and not just on their equilibrium positions. It is a far from simple matter to determine appropriate expressions for these matrix elements and in practice the problem has been tackled rather formally using second quantization procedures.†

Another manifestation of a non-rigid surface is that the intensities of the diffraction peaks are affected by temperature. This phenomenon is well known in the X-ray diffraction from crystals, and is modelled by the Debye–Waller relation, which for our purposes can be written

$$P(mn) = P^0(mn)\exp(-W) \qquad 7.24$$

W is a factor that depends on the temperature average of the displacements of the surface atoms from their equilibrium positions. The most widely used factor for the specular reflection beams, which was proposed by Beeby,††, is

$$W = 4\left(k_z^2 + \frac{2\mu D}{\hbar^2}\right)\langle (r - r^0)^2 \rangle T \qquad 7.25$$

where D is the well depth of the atom–surface potential and $(r - r^0)$ is the displacement of a surface atom in the direction normal to the surface. The mean square displacement from equilibrium is a function of temperature (increasing as T increases). The appearance of the well depth in (7.25) is to allow for the increase in k_z as the atom approaches the surface. Several experiments support the validity of Beeby's formula.

†A. C. Levi, *Nuovo Cimento*, **54B**, 357 (1979).
††J. Beeby, *Comments Solid State Phys.*, **7**, 1 (1975).

Appendix 1

The JWKB approximation

We start with the Schrodinger equation in one dimension which is

$$-\frac{\hbar^2}{2\mu}\frac{d^2\Psi(r)}{dr^2} = [E - V(r)]\Psi(r) \qquad \text{A1.1}$$

Before expanding in powers of $\hbar$ we make the transformation

$$\Psi(r) = e^{iS(r)/\hbar} \qquad \text{A1.2}$$

which on substituting into (A1.1) gives

$$(S')^2 - i\hbar S'' = 2\mu(E - V) \qquad \text{A1.3}$$

The phase function $S(r)$ can now be expanded as follows:

$$S = S_0 + \hbar S_1 + \frac{\hbar^2}{2}S_2 + \cdots \qquad \text{A1.4}$$

and inserting this into (A1.3) we can equate terms of like power in $\hbar$. The terms which are independent of $\hbar$ give the equation

$$(S_0')^2 = 2\mu(E - V) = p^2(r) \qquad \text{A1.5}$$

where $p(r)$ is the momentum. This can be integrated to give

$$S_0 = \pm\int p(r)\,dr \qquad \text{A1.6}$$

The terms from (A1.3) which are linear in $\hbar$ give the equation

$$2S_0'S_1' - iS_0'' = 0 \qquad \text{A1.7}$$

and replacing S_0 using (A1.5), this gives

$$S_1' = \frac{ip'(r)}{2p(r)} \qquad \text{A1.8}$$

which has the solution

$$S_1 = \frac{\mathrm{i}}{2} \ln p(r) = \mathrm{i} \ln [p(r)]^{1/2} \tag{A1.9}$$

To this order the general solution of the equation (A1.1) is a linear combination of the two functions

$$\Psi_{\pm}(r) = \exp\left(\frac{\mathrm{i}}{\hbar}\left\{\pm \int p(r)\,\mathrm{d}r + \mathrm{i}\hbar \ln [p(r)]^{1/2}\right\}\right)$$

$$= \frac{1}{[p(r)]^{1/2}} \exp\left[\pm \frac{\mathrm{i}}{\hbar} \int p(r)\,\mathrm{d}r\right] \tag{A1.10}$$

It is this function that is generally referred to as the JWKB wave function but higher order solutions can be found. For example, the terms of order $\hbar^2$ in (A1.3) are

$$S_0' S_2' + (S_1')^2 - \mathrm{i} S_1'' = 0 \tag{A1.11}$$

and give corrections to the phase factor. Although one cannot define a strict convergence condition for the JWKB expansion one can see from (A1.3) that the first approximation is valid if

$$|\hbar S''| \ll (S')^2 \tag{A1.12}$$

which from (A1.6) requires

$$\hbar |p'(r)| \ll |p^2(r)|$$

or

$$\left|\frac{\mathrm{d}}{\mathrm{d}r}\left[\frac{\hbar}{p(r)}\right]\right| \ll 1 \tag{A1.13}$$

This implies that the gradient of the de Broglie wavelength $\lambda(r) = h/p(r)$ is much less than 2π.

In a scattering experiment the flux of particles must be conserved (assuming particles are not destroyed) and flux is proportional to the velocity of the particles multiplied by the probability density $|\Psi(r)|^2$. It can be seen that the term $p^{-1/2}$ in (A1.10) is necessary to conserve flux.

The difference in phases of the wave function between two points in a classically allowed region $[p(r) > 0]$ is, from (A1.10),

$$\frac{1}{\hbar} \int_{r_1}^{r_2} p(r)\,\mathrm{d}r \tag{A1.14}$$

The integral here is just the classical action for a trajectory leading from (r_1, p_1) to (r_2, p_2). This forms the basis of the postulate in the semi-classical approach to scattering that the classical phase for a trajectory can be used to construct a probability amplitude for a trajectory and thus introduce interference terms into classical theory according to expressions (1.31) and (3.80).

As mentioned in Section 2.4 the JWKB solution fails at a classical turning point $[p(r) = 0]$ which divides a classically accessible from a classically inaccessible region. If a solution, say $\Psi_+(r)$, is known for one side of a classical turning point what is the corresponding solution on the other side? This is known as the connection problem and it has been examined in Section 2.4 for the simplest and most important case of a single isolated turning point. Other, more complicated cases, can be found in M. Child, *Molecular Collision Theory*, Academic Press, 1974.

Appendix 2

Partial wave decomposition of the scattering amplitude

The asymptotic form of the elastic scattering wave function is, from (2.30),

$$\psi \sim e^{ikz} + \frac{f(\theta)e^{ikr}}{r} \qquad \text{A2.1}$$

where $f(\theta)$ is the scattering amplitude. The asymptotic form of the partial wave expansion (2.35) is, using (2.43),

$$\psi \sim \frac{1}{r}\sum_{l=0}^{\infty} A_l P_l(\cos\theta)\sin\left(kr - \frac{l\pi}{2} + \delta_1\right) \qquad \text{A2.2}$$

The coefficients A_l can be obtained by matching these two equations and to do that one must make a partial wave expansion of $\exp(ikz)$ which is[†]

$$e^{ikz} = e^{ikr\cos\theta} = \left(\frac{\pi}{2kr}\right)^{1/2}\sum_{l=0}^{\infty}(2l+1)i^l J_{l+\frac{1}{2}}(kr)P_l(\cos\theta) \qquad \text{A2.3}$$

$J_{l+\frac{1}{2}}(kr)$ being a regular Bessel function. At large r this becomes

$$e^{ikz} \sim \frac{1}{kr}\sum_{l=0}^{\infty}(2l+1)e^{il\pi/2}P_l(\cos\theta)\sin(kr) \qquad \text{A2.4}$$

The scattering amplitude can also be written as an expansion in Legendre polynomials:

$$f(\theta) = \sum_{l=0}^{\infty} f_l P_l(\cos\theta) \qquad \text{A2.5}$$

and substituting (A2.4) and (A2.5) into (A2.1) gives

$$\psi \sim \frac{1}{r}\sum_{l=0}^{\infty} P_l(\cos\theta)\left[\left(\frac{2l+1}{k}\right)e^{i\pi/2}\sin\left(kr - \frac{l\pi}{2}\right) + f_l e^{ikr}\right] \qquad \text{A2.6}$$

[†]M. Abramowitz and I. A. Stegun, *Handbook of Mathematical Functions*, Dover, 1965.

which can now be matched against (A2.2) giving

$$A_l \sin\left(kr - \frac{l\pi}{2} + \delta_1\right) = \left(\frac{2l+1}{k}\right) e^{il\pi/2} \sin\left(kr - \frac{l\pi}{2}\right) + f_l e^{ikr} \qquad \text{A2.7}$$

Writing both sine functions as exponentials we can now equate the coefficients of $\exp(ikr)$ in the expansion of (A2.1) and (A2.2) to obtain

$$\frac{1}{2i} A_l e^{-il\pi/2 + i\delta_l} = \frac{2l+1}{2ik} + f_l \qquad \text{A2.8}$$

Likewise, for the coefficients of $\exp(-ikr)$ we get

$$\frac{1}{2i} A_l e^{il\pi/2 - i\delta_l} = \left(\frac{2l+1}{2ik}\right) e^{il\pi} \qquad \text{A2.9}$$

From these two equations we eliminate A_l and obtain

$$f_l = \left(\frac{2l+1}{2ik}\right)(e^{2i\delta_l} - 1) \qquad \text{A2.10}$$

which gives the scattering amplitude as (2.51):

$$f(\theta) = \frac{1}{2ik} \sum_{l=0}^{\infty} (2l+1)(e^{2i\delta_l} - 1) P_l(\cos\theta) \qquad \text{A2.11}$$

Appendix 3

The use of Jost functions to calculate the **S** *matrix for inelastic scattering*

The multichannel equations in the scattering coordinate r must be integrated outwards from $r = 0$ or a sufficiently small value of r that the wave function in all channels can be taken as zero. One method for doing this, leading to what are called Jost functions, has been described in Section 3.5. The Jost functions A_{ij} and B_{ij} are coefficients in the asymptotic forms of the regular solutions of the radial wave equation (e.g. (4.50) or the collinear form (3.69)) which have the form

$$\mathbf{\Psi}(r) \sim e^{i\mathbf{k}r}\mathbf{A} + e^{-i\mathbf{k}r}\mathbf{B} \qquad \text{A3 1}$$

where $\mathbf{k}$ is a diagonal matrix of the wave vectors k_i. For three-dimensional scattering $\exp(i\mathbf{k}r)$ is replaced by $\exp(i\mathbf{k}r - l\pi/2)$.

A and **B** are square matrices whose dimension equals the number of channels, and these channels can be open or closed. To obtain the **S** matrix one must transform (A3.1) so that the coefficients of exponentially growing terms are zero for closed channels. A closed channel has imaginary k_i (conventionally positive imaginary) and therefore the closed channel elements of **B** must be transformed to zero.

We start by partitioning the Jost matrices into open and closed blocks in an obvious way:

$$\mathbf{A} = \begin{vmatrix} \mathbf{A}_{OO} & \mathbf{A}_{OC} \\ \mathbf{A}_{CO} & \mathbf{A}_{CC} \end{vmatrix}, \quad \mathbf{B} = \begin{vmatrix} \mathbf{B}_{OO} & \mathbf{B}_{OC} \\ \mathbf{B}_{CO} & \mathbf{B}_{CC} \end{vmatrix} \qquad \text{A3.2}$$

For the **B** matrix a transformation can easily be found which leaves only the open–open elements non-zero:

$$\begin{bmatrix} \mathbf{B}_{OO} & \mathbf{B}_{OC} \\ \mathbf{B}_{CO} & \mathbf{B}_{CC} \end{bmatrix} \begin{bmatrix} \mathbf{I} & 0 \\ \mathbf{\Gamma} & 0 \end{bmatrix} = \begin{bmatrix} \mathbf{B}'_{OO} & 0 \\ 0 & 0 \end{bmatrix} \qquad \text{A3.3}$$

which requires

$$\mathbf{B}_{CO} + \mathbf{B}_{CC}\mathbf{\Gamma} = 0$$

or $\qquad\qquad$ A3.4

$$\mathbf{\Gamma} = -\mathbf{B}_{CC}^{-1}\mathbf{B}_{CO}$$

and for this

$$\mathbf{B}'_{OO} = \mathbf{B}_{OO} - \mathbf{B}_{OC}\mathbf{B}_{CC}^{-1}\mathbf{B}_{CO} \qquad \text{A3.5}$$

The same transformation on **A** gives an open–open submatrix which is

$$\mathbf{A}'_{OO} = \mathbf{A}_{OO} - \mathbf{A}_{OC}\mathbf{B}_{CC}^{-1}\mathbf{B}_{CO} \qquad \text{A3.6}$$

Although we have now ensured that the submatrices of **B** are zero for closed channels, $\mathbf{A}'_{OO}$ is not diagonal which implies that the incoming waves are not in pure internal states. We therefore make a further transformation to make the coefficients of e^{ikr} unity, which involves a pre-multiplication of $\mathbf{A}'_{OO}$ and $\mathbf{B}'_{OO}$ by $(\mathbf{A}'_{OO})^{-1}$. The resulting **B** matrix for the open channels is therefore

$$\begin{aligned}\mathbf{B}''_{OO} &= \mathbf{B}'_{OO}(\mathbf{A}'_{OO})^{-1}\\ &= (\mathbf{B}_{OO} - \mathbf{B}_{OC}\mathbf{B}_{CC}^{-1}\mathbf{B}_{CO})(\mathbf{A}_{OO} - \mathbf{A}_{OC}\mathbf{B}_{CC}^{-1}\mathbf{B}_{CO})^{-1}\end{aligned} \qquad \text{A3.7}$$

The asymptotic form of the wave functions for the open channels is therefore of the form

$$\boldsymbol{\Psi}(r) \sim e^{i\mathbf{k}r} + e^{-i\mathbf{k}r}\mathbf{B}''_{OO} \qquad \text{A3.8}$$

and comparison with (4.35) and (4.36) leads to **S**-matrix elements of the form

$$S_{ji} = -\left(\frac{k_i}{k_j}\right)^{1/2} B''_{ji} \qquad \text{A3.9}$$

Appendix 4

Numerical methods

The two main numerical problems that have to be faced in scattering theory are the solution of the multichannel equations and the calculation of trajectories from Hamilton's equations. For real potentials neither of these problems can be solved analytically. In both cases we are dealing with coupled differential equations—essentially second-order equations, but there is an easy transformation to a set of first-order equations (e.g. the conversions of Newton's equations to Hamilton's equations).

The multichannel equations have solutions that are rapidly oscillating in the classically allowed regions but have exponential growth in the classically forbidden regions. It is the combination of these two types of behaviour that makes the choice of numerical technique important. The solutions of Hamilton's equations are not generally rapidly oscillating but, as described in Section 5.2, the equations are generally stiff.

Nearly all numerical methods for integrating differential equations extrapolate the function by a Taylor expansion to a certain order. Broadly speaking there is a choice between single-step methods, where the step length is varied, or multistep methods, where the step length is usually constant. In the former are the Runge–Kutta methods, which for trajectories are generally very efficient, particularly when the program possesses both a variable-step length and a variable-order routine. Many computer packages are available for these methods and their relative merits for trajectory problems have been investigated.† However, there is no best package for all problems and methods are continually being improved, so that there is no substitute for an initial examination of a particular problem with more than one routine.

The above methods can be applied to any type of differential equation. The multichannel equations are special in that they are linear and for these more efficient methods are available. The equations have the general form

$$\boldsymbol{\psi}(r)'' = \mathbf{F}(r)\boldsymbol{\psi}(r) \qquad \text{A4.1}$$

†P. Brunner, *J. Comp. Phys.*, **14**, 391 (1974).

where $\mathbf{F}(r)$ is an $n \times n$ symmetric matrix. If the eigenvalues of $F(r)$ are all negative then one is in a classically allowed region for all channels and there is usually no difficulty with the integration. For this case, and particularly for single-channel elastic scattering, we recommend the Numerov method.

A Taylor expansion of $\psi_\pm(r \pm \Delta)$ generates the function

$$\psi_+ + \psi_- = 2\psi(r) + \Delta^2\psi''(r) + \frac{\Delta^4}{12}\psi''''(r) + \cdots \qquad \text{A4.2}$$

Differentiating this expression twice, and with (A4.2) eliminating fourth derivatives of $\psi(r)$, leads to the expression

$$\psi_+ + \psi_- - \frac{\Delta^2}{12}(\psi''_+ + \psi''_-) = 2\psi + \frac{5\Delta^2}{6}\psi'' + O(\Delta^6) \qquad \text{A4.3}$$

Making use of (A4.1) to remove second derivatives and neglecting terms $O(\Delta^6)$ gives the expression

$$\psi_+ = \left[\mathbf{I} - \frac{\Delta^2}{12}\mathbf{F}(r+\Delta)\right]^{-1}\left\{\left[2\mathbf{I} + \frac{5\Delta^2}{6}\mathbf{F}(r)\right]\psi - \left[\mathbf{I} - \frac{\Delta^2}{12}\mathbf{F}(r-\Delta)\right]\psi_-\right\} \qquad \text{A4.4}$$

which is called the Numerov formula. Initial values for a regular solution can be taken as $\psi = 0$ and $\psi' = \mathbf{I}$.

When one or more of the eigenvalues of $F(r)$ are positive, direct integration of (A.4.1) is very difficult because some of the components of ψ grow exponentially as r increases. In such a case the integration method most commonly used is the log derivative method.

By analogy with the expression

$$\frac{\mathrm{d}\ln y}{\mathrm{d}x} = \frac{1}{y}\frac{\mathrm{d}y}{\mathrm{d}x} \qquad \text{A4.5}$$

we define the log derivative of the matrix ψ by

$$\mathbf{X} = \psi'\psi^{-1} \qquad \text{A4.6}$$

Taking the derivatives of this we have

$$\mathbf{X}' = \psi''\psi^{-1} - \psi'\psi^{-1}\psi'\psi^{-1} \qquad \text{A4.7}$$

which from (A4.1) and (A4.6) gives a set of first-order non-linear equations for $\mathbf{X}$:

$$\mathbf{X}' = \mathbf{F} - \mathbf{X}^2 \qquad \text{A4.8}$$

A convenient algorithm for integrating this equation was derived by Johnson.† If X_n is the value of X at $r = r_0 + nh$, then

$$\mathbf{X}_n = (\mathbf{I} + h\mathbf{X}_{n-1})^{-1}\mathbf{X}_{n-1} - \frac{h}{3}W_n\mathbf{G}_n \qquad \text{A4.9}$$

†B. R. Johnson, *J. Comp. Phys.* **13**, 445 (1973).

where h is the step length and $\mathbf{G}_n$ is given by

$$\mathbf{G}_n = \begin{cases} -\mathbf{F}_n, & n = 0, 2, 4, \ldots, N \\ -\left(\mathbf{I} - \dfrac{h^2}{6}\mathbf{F}_n\right)^{-1}\mathbf{F}_n, & n = 1, 3, 5, \ldots, N-1 \end{cases} \qquad \text{A4.10}$$

N is the total number (even) of integration points and W_n are weights given by

$$W_n = \begin{cases} 1, & n = 0, N \\ 4, & n = 1, 3, 5, \ldots, N-1 \\ 2, & n = 2, 4, 6, \ldots, N-2 \end{cases} \qquad \text{A4.11}$$

Integration begins at a point r_0 where $\boldsymbol{\psi} = \mathbf{O}$ and from (A.6) this implies $\mathbf{X}_0 = \infty$; for computational purposes one takes, say, $\mathbf{X}_0 = \mathbf{I} \times 10^{20}$.

Having $\mathbf{X}$ we determine the $\mathbf{S}$-matrix by examining the wave function in the asymptotic limit where the potential is zero. In this region the wave function can be expressed in terms of Jost functions (A3.1):

$$\boldsymbol{\psi} \sim \mathrm{e}^{\mathrm{i}\mathbf{k}r}\mathbf{A} + \mathrm{e}^{-\mathrm{i}\mathbf{k}r}\mathbf{B} \qquad \text{A4.12}$$

and its derivative is

$$\boldsymbol{\psi}' = \mathrm{i}\mathbf{k}(\mathrm{e}^{\mathrm{i}\mathbf{k}r}\mathbf{A} - \mathrm{e}^{-\mathrm{i}\mathbf{k}r}\mathbf{B}) \qquad \text{A4.13}$$

From these equations we obtain $\mathbf{A}$ and $\mathbf{B}$ in the form

$$\begin{aligned} 2\mathbf{A} &= \mathrm{e}^{-\mathrm{i}\mathbf{k}r}(\boldsymbol{\psi} - \mathrm{i}\mathbf{k}^{-1}\boldsymbol{\psi}') \\ 2\mathbf{B} &= \mathrm{e}^{\mathrm{i}\mathbf{k}r}(\boldsymbol{\psi} + \mathrm{i}\mathbf{k}^{-1}\boldsymbol{\psi}') \end{aligned} \qquad \text{A4.14}$$

or alternatively

$$\begin{aligned} 2\mathbf{A} &= -\mathrm{i}\mathrm{e}^{-\mathrm{i}\mathbf{k}r}\mathbf{k}^{-1}(\mathrm{i}\mathbf{k} + \mathbf{X})\boldsymbol{\psi} \\ 2\mathbf{B} &= \mathrm{i}\mathrm{e}^{+\mathrm{i}\mathbf{k}r}\mathbf{k}^{-1}(-\mathrm{i}\mathbf{k} + \mathbf{X})\boldsymbol{\psi} \end{aligned} \qquad \text{A4.15}$$

Although these expressions contain functions with an explicit r dependence, it can be shown that they are nevertheless independent of r. The $\mathbf{S}$ matrix is given by (3.97), and if we use (A4.15) we obtain

$$\mathbf{S} = \mathbf{k}^{1/2}\mathbf{A}\mathbf{B}^{-1}\mathbf{k}^{-1/2} = -\mathrm{e}^{-\mathrm{i}\mathbf{k}r}\mathbf{k}^{-1/2}(\mathrm{i}\mathbf{k} + \mathbf{X})(-\mathrm{i}\mathbf{k} + \mathbf{X})^{-1}\mathbf{k}^{1/2}\mathrm{e}^{-\mathrm{i}\mathbf{k}r} \qquad \text{A4.16}$$

which depends only on the asymptotic limit of the log derivative matrix. If some of the channels are closed then the Jost matrices (A4.15) should be modified according to the prescription in Appendix 3, prior to obtaining (A4.16).

Appendix 5

Effective impact parameter for a three-dimensional surface

We consider a surface for which there is only one value of r for each θ and ϕ, and this is given by the function

$$r = f(\theta, \phi) \tag{A5.1}$$

The normal vector to the surface is given by

$$\mathbf{n} = \frac{\boldsymbol{\nabla} F}{|\boldsymbol{\nabla}\mathbf{F}|} \tag{A5.2}$$

where

$$F = r - f(\theta, \phi) \tag{A5.3}$$

To define the impact parameter for a hard shape collision we must first define the origin of r to be the centre of mass of the molecule. The effective impact parameter b_n is then the perpendicular distance from the origin to $\mathbf{n}$ which is

$$b_n = |\mathbf{n} \times \mathbf{r}| = \frac{|\boldsymbol{\nabla} F \times \mathbf{r}|}{|\boldsymbol{\nabla}\mathbf{F}|} \tag{A5.4}$$

The gradient operator in polar coordinates is

$$\nabla = \mathbf{r}_0 \frac{\partial}{\partial r} + \frac{\boldsymbol{\theta}_0}{r}\frac{\partial}{\partial \theta} + \frac{\boldsymbol{\phi}_0}{r \sin\theta}\frac{\partial}{\partial \phi} \tag{A5.5}$$

so that

$$\boldsymbol{\nabla} F \times \mathbf{r} = r \boldsymbol{\nabla} F \times \mathbf{r}_0 = r \frac{\partial F}{\partial \theta} \boldsymbol{\theta}_0 \times \mathbf{r}_0 + \frac{1}{\sin\theta}\frac{\partial F}{\partial \phi} \boldsymbol{\phi}_0 \times \mathbf{r}_0 \tag{A5.6}$$

From (A5.3) and (A5.4) we obtain the result

$$b_n = \frac{\left[\left(\frac{\partial f}{\partial \theta}\right)^2 + \frac{1}{\sin^2\theta}\left(\frac{\partial f}{\partial \phi}\right)^2\right]^{1/2}}{\left[1 + \frac{1}{r^2}\left(\frac{\partial f}{\partial \theta}\right)^2 + \frac{1}{r^2 \sin^2\theta}\left(\frac{\partial f}{\partial \phi}\right)^2\right]} \tag{A5.7}$$

The extrema of b_n are defined by

$$\frac{\partial b_n}{\partial \theta} = 0; \qquad \frac{\partial b_n}{\partial \phi} = 0 \qquad \text{A5.8}$$

and there will be lines on the surface $f(\theta, \phi)$ corresponding to these extrema.

Appendix 6

Hamilton's equations for A + BC classical trajectory calculations

The space-fixed coordinates of the atoms are x_A, y_A, z_A, etc., and their conjugate momenta $p_{x_A}, p_{y_A}, p_{z_A}$, etc. These are transformed to give coordinates of C relative to B as origin:

$$Q_1 = x_C - x_B, \qquad Q_2 = y_C - y_B, \qquad Q_3 = z_C - z_B \qquad \text{A6.1}$$

and coordinates of A relative to the centre of mass of BC:

$$Q_4 = x_A - (m_B x_B + m_C x_C)(m_B + m_C)^{-1} \qquad \text{A6.2}$$

with similar expressions for Q_5 and Q_6 related to the y and z values respectively. The coordinates of the centre of mass of the whole are

$$Q_7 = \frac{m_A x_A + m_B x_B + m_C x_C}{m_A + m_B + m_C} \qquad \text{A6.3}$$

with similar expressions for the y and z components Q_8 and Q_9.

The momenta conjugate to the Q_i are determined from the transformation

$$p_i = \sum_j P_j \frac{\partial Q_j}{\partial q_i} \qquad \text{A6.4}$$

and the kinetic energy of the system in terms of these momenta is

$$2T = \mu_{BC}^{-1} \sum_{i=1}^{3} P_i^2 + \mu_{A,BC}^{-1} \sum_{i=4}^{6} P_i^2 + M^{-1} \sum_{i=7}^{9} P_i^2 \qquad \text{A6.5}$$

where

$$\mu_{BC} = \frac{m_B m_C}{m_B + m_C}$$

$$\mu_{A,BC} = \frac{m_A(m_B + m_C)}{M} \qquad \text{A6.6}$$

$$M = m_A + m_B + m_C$$

The potential energy V is independent of the position of the centres of mass and hence depends only on $Q_1, \ldots, Q_6$. It is usually expressed in terms of interatomic distances

$$\begin{aligned} R_{AB} &= [(x_B - x_A)^2 + (y_B - y_A)^2 + (z_B - z_A)^2]^{1/2} \\ &= \{[m_C(m_B + m_C)^{-1}Q_1 + Q_4]^2 + [m_C(m_B + m_C)^{-1}Q_2 + Q_5]^2 \\ &\quad + [m_C(m_B + m_C)^{-1}Q_3 + Q_6]^2\}^{1/2}, \qquad \text{etc.} \end{aligned} \tag{A6.7}$$

and the derivatives of V with respect to the Q_i are obtained by the chain rule

$$\frac{\partial V}{\partial Q_i} = \sum_j \left(\frac{\partial V}{\partial R_j}\right)\left(\frac{\partial R_j}{\partial Q_i}\right) \tag{A6.8}$$

where $R_j = R_{AB}$, R_{BC} and R_{AC}.

Hamilton's equations are

$$\dot{Q}_i = \left(\frac{\partial H}{\partial P_i}\right), \qquad \dot{P}_i = -\left(\frac{\partial H}{\partial Q_i}\right) \tag{A6.9}$$

The momenta of the centre of mass are constants of the motion and need not be considered further. From (A6.4) it can be seen that the time derivatives of the $\dot{Q}_i$ are given by the simple expressions

$$\dot{Q}_i = \mu_{BC}^{-1} P_i \qquad (i = 1\text{–}3) \tag{A6.10}$$

$$\dot{Q}_i = \mu_{A,BC}^{-1} P_i \qquad (i = 4\text{–}6) \tag{A6.11}$$

The equations for $\dot{P}_i$ are more complicated and depend on the derivatives of the potential evaluated through (A6.7):

$$\begin{aligned} -\dot{P}_i = {} & \frac{m_C}{m_B + m_C}\left(\frac{m_C}{m_B + m_C} Q_i + Q_{i+3}\right)\frac{1}{R_{AB}}\frac{\partial V}{\partial R_{AB}} \\ & + \frac{m_B}{m_B + m_C}\left(\frac{m_B}{m_B + m_C} Q_i - Q_{i+3}\right)\frac{1}{R_{AC}}\frac{\partial V}{\partial R_{AC}} \\ & + \frac{Q_i}{R_{BC}}\frac{\partial V}{\partial R_{BC}} \qquad (i = 1\text{–}3) \end{aligned} \tag{A6.12}$$

$$-\dot{P}_i = \left(\frac{m_C}{m_B + m_C} Q_i + Q_{i+3}\right)\frac{1}{R_{AB}}\frac{\partial V}{\partial R_{AB}} - \left(\frac{m_B}{m_B + m_C} Q_i - Q_{i+3}\right)\frac{1}{R_{AC}}\frac{\partial V}{\partial R_{AC}} \qquad (i = 4\text{–}6) \tag{A6.13}$$

The initial values of these variables are determined according to the arguments given in Section 5.2. The diatomic internuclear distance is put equal to one of its classical turning points in the appropriate (v, j) state, $R_\pm$. These are determined by solving the equation

$$\frac{\hbar^2 j(j+1)}{2\mu_{BC} R^2} + V(R) = E_{vj} \tag{A6.14}$$

where $V(R)$ is the diatomic potential which may, for example, be represented by a Morse function. This equation can be solved by Newton's method. The calculations proceed by selecting R_+ or R_- randomly.

From Figure 5.5 it can be seen that the initial values of Q_1, Q_2 and Q_3 are given by

$$Q_1^0 = R_\pm \sin\theta\cos\phi \tag{A6.15}$$

$$Q_2^0 = R_\pm \sin\theta\sin\phi \tag{A6.16}$$

$$Q_3^0 = R_\pm \cos\theta \tag{A6.17}$$

The initial values of the conjugate momenta depend also on the angle η that the angular momentum operator makes with a normal to the BC bond which we can take in the direction of the vector $\mathbf{R}_{BC} \times \mathbf{Z}$.

$$P_1^0 = J(\sin\phi\cos\eta - \cos\theta\cos\phi\sin\eta)R_\pm^{-1} \tag{A6.18}$$

$$P_2^0 = -J(\cos\phi\cos\eta + \cos\theta\sin\phi\sin\eta)R_\pm^{-1} \tag{A6.19}$$

$$P_3^0 = J(\sin\theta\sin\eta)R_\pm^{-1} \tag{A6.20}$$

where J is the rotational angular momentum which is related to the j quantum number by

$$J = \hbar[j(j+1)]^{1/2} \tag{A6.21}$$

The initial values of the relative coordinates of A to the centre of mass of BC are

$$Q_4^0 = Q_5^0 = 0 \tag{A6.22}$$

$$Q_6^0 = -(d + \zeta\tau v_r) \tag{A6.23}$$

where d is a distance greater than the range of the A–BC potential, ζ is a random number between 0 and 1, τ the vibrational period of BC and v_r the relative velocity of A to BC. The conjugate momenta are

$$P_4^0 = P_5^0 = 0 \tag{A6.24}$$

$$P_6^0 = \mu_{A,BC} v_r \tag{A6.25}$$

As well as the random number to choose the initial distance of A from BC, one has to generate random numbers for the angles θ, ϕ and η. Both ϕ and η are chosen randomly in the range 0 to 2π, but θ, which has a range 0 to π, must be chosen to give equal weighting to each element of solid angle:

$$d\omega = \sin\theta\, d\theta\, d\phi \tag{A6.26}$$

This requirement is satisfied by selecting $\cos\theta$ randomly between -1 and $+1$.

As explained in the text, the initial b value is generally selected uniformly from 0 to b_{max} and the trajectory is weighted by its b value. For product analysis one must first determine the nature of the products by examining the asymptotic form of Q_i and determining from these the three internuclear distances. The transformations to give the rotational and vibrational energy of the diatomic, the relative velocity vector and the scattering angle will depend on which of the three possible outcomes occurs, but all follow the same type of analysis used in setting up the equations for the reactants.

Bibliography

The bibliography contains general references which we have found most valuable for our studies in scattering theory and, particularly, in the writing of this book. It is divided according to the chapters of greatest relevance, although many works are valuable for more than one chapter. We pick out three books that cover very large parts of the field; they are quite different in character and, we believe, complementary to our own.

M. S. Child, *Molecular Collision Theory*, Academic Press, London and New York, 1974. This is an excellent theory book, particularly strong on semi-classical methods.

R. D. Levine and R. B. Bernstein, *Molecular Reaction Dynamics and Chemical Reactivity*, Oxford University Press, New York, Oxford, 1987. This is an excellent book for the link between theory and experiment.

R. G. Newton, *Scattering Theory of Waves and Particles*, McGraw-Hill, New York, 1966. This is an advanced theory book which deals particularly well with formal scattering theory.

Chapter 1

R. B. Bernstein, Introduction to atom–molecule collisions: the interdependency of theory and experiment, in R. B. Bernstein (ed.), *Atom–Molecule Collision Theory. A Guide for the Experimentalist*, Plenum Press, New York, 1979.

U. Buck, Elastic scattering, *Adv. Chem. Phys.*, **30**, 313 (1975).

E. E. Nikitin, *Theory of Elementary Atomic and Molecular Processes in Gases*, Clarendon Press, Oxford, 1974.

J. P. Toennies, Molecular Beam scattering experiments on elastic, inelastic, and reactive collisions, in H. Eyring, W. Jost and D. Henderson (eds.), *Physical Chemistry*, Vol. 6, Academic Press, New York, 1974–5.

Chapter 2

M. V. Berry and K. E. Mount, Semi-classical approximations in wave mechanics, *Rep. Prog. Phys.*, **35**, 315 (1972).

D. Bohm, *Quantum Theory*, Prentice-Hall, London, 1960.

H. Goldstein, *Classical Mechanics*, Addison-Wesley, New York, 1950.

R. E. Johnson, *Introduction to Atomic and Molecular Collisions*, Plenum Press, New York, 1982.

K. P. Lawley (ed.), Molecular beam scattering, *Adv. Chem. Phys.*, **30**, 1975.

H. S. W. Massey, *Atomic and Molecular Collisions*, Taylor and Francis, London, 1979.
N. F. Mott and H. S. W. Massey, *The Theory of Atomic Collisions*, Clarendon Press, Oxford, 1965.
H. Pauly, Elastic scattering cross-sections, in R. B. Bernstein (ed.), *Atom–Molecule Collision Theory. A Guide to the Experimentalist*, Plenum Press, New York, 1979.

Chapter 3

A. E. De Presto and H. Rabitz, Vibrational and rotational collision processes, *Adv. Chem. Phys.*, **42**, 271 (1980).
A. S. Dickinson, Non-reactive heavy particle collision calculations, *Comp. Phys. Commun.*, **17**, 51 (1979).
G. A. Fisk and F. F. Crim, Single collision studies of vibrational energy transfer mechanics, *Acc. Chem. Res.*, **10**, 73 (1977).
W. R. Gentry, Vibrational excitation: classical and semi-classical methods, in R. B. Bernstein (ed.), *Atom–Molecule Collision Theory. A Guide for the Experimentalist*, Plenum Press, New York, 1979.
D. Rapp and T. Kassal, The theory of vibrational energy transfer between simple molecules in non-reactive collisions, *Chem. Rev.*, **69**, 61 (1969).
D. Secrest, Theory of rotational and vibrational energy transfer in molecules, *Ann. Rev. Phys. Chem.*, **24**, 379 (1973).
H. K. Shin, Vibrational energy transfer, in W. H. Miller (ed.), *Dynamics of Molecular Collisions*, Plenum Press, New York, 1976.

Chapter 4

S. D. Bosanac, *Long-Lived States in Collisions*, CRC Press, Boca Raton, Fla., 1988.
J. M. Bowman (ed.), *Molecular Collision Dynamics*, Springer-Verlag, Berlin, 1983.
T. A. Brunner and D. Pritchard, Fitting laws for rotationally inelastic collisions, *Adv. Chem. Phys.*, **50**, 589 (1982).
U. Buck, Rotationally inelastic scattering of hydrogen molecules and the non-spherical interaction, *Faraday Disc. Chem. Soc.*, **73**, 187 (1982).
M. Faubel and J. P. Toennies, Scattering studies of rotational and vibrational excitation in molecules, *Adv. At. Molec. Phys.*, **13**, 229 (1977).
D. J. Kouri, Rotational excitation, in R. B. Bernstein (ed.), *Atom–Molecule Collision Theory. A Guide to the Experimentalist*, Plenum Press, New York, 1979.
R. J. Le Roy, G. C. Corey and J. D. Hutson, Predissociation of weak-anisotropy, van der Waals molecules, *Faraday Disc. Chem. Soc.*, **73**, 339 (1982).
H. J. Loesch, Scattering of non-spherical molecules, *Adv. Chem. Phys.*, **42**, 421 (1980).
A. J. McCaffery, M. J. Proctor and B. J. Whitaker, Rotational energy transfer: polarization and scaling, *Ann. Rev. Phys. Chem.*, **37**, 223 (1986).
J. P. Toennies, The calculation and measurement of cross-sections for rotational and vibrational excitation, *Ann. Rev. Phys. Chem.*, **27**, 225 (1976).

Chapter 5

M. Baer, *The Theory of Chemical Reaction Dynamics*, CRC Press, Boca Raton, Fla., 1985.
D. C. Clary (ed), *The Theory of Chemical Reaction Dynamics*, Reidel, Boston, 1986.
K. J. Landler and M. C. King, The development of transition state theory, *J. Phys. Chem.*, **87**, 2657 (1983).

R. A. Marcus, Statistical theory of unimolecular reactions and intramolecular dynamics, *Laser Chem.*, **2**, 203 (1983).
W. H. Miller, Reaction path models for polyatomic reaction dynamics—from transition state theory to path integrals, in D. C. Clary (ed.), *The Theory of Chemical Reaction Dynamics*, Reidel, Boston, 1986.
J. T. Muckerman, Applications of classical trajectory techniques to reactive scattering, in D. Henderson (ed.), *Theoretical Chemistry*, Academic Press, New York, 1981.
J. N. Murrell, S. Carter, S. C. Farantos, P. Huxley and A. J. C. Varandas, *Molecular Potential Energy Functions*, John Wiley and Sons, Chichester, 1984.
R. N. Porter, Molecular trajectory calculations, *Ann. Rev. Phys. Chem.*, **25**, 317 (1974).
L. M. Raff and D. L. Thompson, Classical trajectory approach to reactive scattering, in M. Baer (ed.), *The Theory of Chemical Reaction Dynamics*, CRC Press, Boca Raton, Fla., 1985.
G. C. Schatz, Overview of reactive scattering, in D. G. Truhler (ed.), *Potential Energy Surfaces and Dynamics Calculations*, Plenum Press, New York, 1981.
G. C. Schatz, Tunnelling in bimolecular collisions, *Chem. Rev.*, **87**, 81 (1987).
D. G. Truhlar and B. C. Garrett, Variational transition state theory, *Ann. Rev. Phys. Chem.*, **35**, 159 (1984).
D. G. Truhlar, R. Steckler and M. S. Gordon, Potential energy surfaces for polyatomic reaction dynamics, *Chem. Rev.*, **87**, 217 (1987).

Chapter 6

M. Baer, Theory of electronic nonadiabatic transitions in chemical reactions, in M. Baer (ed.), *The Theory of Chemical Reaction Dynamics*, CRC Press, Boca Raton, Fla., 1985.
M. S. Child, Electronic excitation, in R. B. Bernstein (ed.), *Atom–Molecule Collision Theory. A Guide for the Experimentalist*, Plenum Press, New York, 1979.
P. J. Dagdigian and M. L. Campbell, Spin–orbit effects in gas phase chemical reactions, *Chem. Rev.*, **87**, 1 (1987).
I. V. Hertel, Progress in electronic to vibrational energy transfer, *Adv. Chem. Phys.*, **50**, 475 (1982).
J. Jortner and S. Mukamel, Radiationless transitions, *Int. Rev. Sci. Phys. Chem.*, **1**, 329 (1975).
H. Nakamura, Unified treatment of nonadiabatic transitions in the rotating frame of the complex, in J. Eichler, I. V. Hertel and N. Stolterfot (eds.), *Electronic and Atomic Collisions*, North-Holland, Amsterdam, 1984.
E. E. Nikitin and S. Ya. Umanskii, *Theory of Slow Atomic Collisions*, Springer-Verlag, Berlin, 1984.
J. C. Tully, Nonadiabatic processes in molecular collisions, in W. H. Miller (ed.), *Dynamics of Molecular Collisions*, Plenum Press, New York, 1976.
L. Zulicke, Ch. Zuhrt and S. Ya. Umansky, Some problems of the dynamics of nonadiabatic energy processes, in J. Eichler, I. V. Hertel and N. Stolterfot (eds.), *Electronic and Atomic Collisions*, North-Holland, Amsterdam, 1984.

Chapter 7

G. Benedek and U. Valbusa (eds.), *Dynamics of Gas–Surface Interaction*, Springer-Verlag, Berlin, 1982.
S. L. Bernasek, State-resolved dynamics of chemical reactions at surfaces, *Chem. Rev.*, **87**, 91 (1987).
R. B. Gerber, Molecular scattering from surfaces: theoretical methods and results, *Chem Rev.*, **87**, 29 (1987).

B. Pullman, J. Jortner, A. Nitzan and R. B. Gerber (eds.), *Dynamics on Surfaces* Reidel, Boston, 1984.

J. C. Tully, Theories of the dynamics of inelastic and reactive processes at surfaces, *Ann. Rev. Phys. Chem.*, **31**, 319 (1980).

G. Wolken, The scattering of atoms and molecules from solid surfaces, in W. H. Miller (ed.), *Dynamics of Molecular Collisions*, Plenum Press, New York, 1976.

Index

Index of systems cited